DATE DUE

NOV 17			

THE DESIGN AND IMPLEMENTATION OF LOW-POWER CMOS RADIO RECEIVERS

THE DESIGN AND IMPLEMENTATION OF LOW-POWER CMOS RADIO RECEIVERS

Derek K. Shaeffer
Stanford University

Thomas H. Lee
Stanford University

KLUWER ACADEMIC PUBLISHERS
Boston / Dordrecht / London

Distributors for North, Central and South America:
Kluwer Academic Publishers
101 Philip Drive, Assinippi Park
Norwell, Massachusetts 02061 USA
Telephone (781) 871-6600
Fax (781) 871-6528
E-Mail <kluwer@wkap.com>

Distributors for all other countries:
Kluwer Academic Publishers Group
Distribution Centre
Post Office Box 322
3300 AH Dordrecht, THE NETHERLANDS
Telephone 31 78 6392 392
Fax 31 78 6546 474
E-Mail <orderdept@wkap.nl>

Electronic Services <http://www.wkap.nl>

Library of Congress Cataloging-in-Publication Data
Shaeffer, Derek K.
The design and implementation of low-power CMOS radio receivers / Derek K. Shaeffer, Thomas H. Lee
p. cm.
Includes bibliographical references and index.
ISBN 0-7923-8518-7 (alk. paper)
1. Radio--Receivers and reception--Design and construction.
2. Low voltage integrated circuits--Design and construction.
3. Metal oxide semiconductors. Complementary--Design and construction. I. Lee, Thomas H., 1959- . II. Title.
TK6563.S412 1999
621.384'18--dc21 99-20467
CIP

Printed on acid-free paper.

Printed in the United States of America

To all of our teachers, and to our parents, who were the best teachers of them all.

Contents

List of Figures

List of Tables

Foreword

It is hardly a profound observation to note that we remain in the midst of a wireless revolution. In 1998 alone, over 150 million cell phones were sold worldwide, representing an astonishing 50% increase over the previous year. Maintaining such a remarkable growth rate requires constant innovation to decrease cost while increasing performance and functionality.

Traditionally, wireless products have depended on a mixture of semiconductor technologies, spanning GaAs, bipolar and BiCMOS, just to name a few. A question that has been hotly debated is whether CMOS could ever be suitable for RF applications. However, given the acknowledged inferiority of CMOS transistors relative to those in other candidate technologies, it has been argued by many that "CMOS RF" is an oxymoron, an endeavor best left cloistered in the ivory towers of academia.

In rebuttal, there are several compelling reasons to consider CMOS for wireless applications. Aside from the exponential device and density improvements delivered regularly by Moore's law, only CMOS offers a technology path for integrating RF and digital elements, potentially leading to exceptionally compact and low-cost devices. To enable this achievement, several thorny issues need to be resolved. Among these are the problem of poor passive components, broadband noise in MOSFETs, and phase noise in oscillators made with CMOS. Beyond the component level, there is also the important question of whether there are different architectural choices that one would make if CMOS were used, given the different constraints.

The work described in this book, based on Dr. Shaeffer's doctoral research at Stanford, is a significant first step toward answering many of these questions. This single-chip GPS receiver actually outperforms existing implementations in other technologies, while consuming less power. Furthermore, it is more highly integrated. As is made apparent in the chapters to come, this performance is made possible by a careful choice of architecture, and a detailed study of how to approach performance limits consistently. Important advances in understanding how to design low-noise amplifiers (LNAs) and wide dynamic range filters in CMOS form the core contributions of this work. Just as important are the scaling properties elucidated by this research, for it makes it clear that both RF and digital performance will improve together, assuring that CMOS will become an important medium in which to realize RF circuits and systems.

Thomas H. Lee
Stanford University

Acknowledgments

We would like to thank the many friends and colleagues who have contributed to this work. In particular, we thank Professors Bruce Wooley and Donald Cox whose thoughtful comments strengthened the final manuscript. We are also indebted to the members of the Stanford Microwave Integrated Circuits group (SMIrC), including Dr. Arvin Shahani, Dr. Ali Hajimiri, Dave Colleran, Hamid Rategh, Hirad Samavati, Kevin Yu, Mar Hershenson, Sunderarajan Mohan, Rafael Betancourt, Ramin Farjad-Rad and Tamara Ahrens. The SMIrC group has been an incredibly stimulating and enjoyable group to work with, and we thank the members for their enthusiasm and their technical excellence. In particular, we thank Arvin, Hamid, Hirad, Mar, and Mohan who worked so diligently on the GPS receiver project (code named "Waldo") along with Dr. Patrick Yue, Min Xu and Dan Eddleman.

We would like to thank the many members of the Wooley, Wong, Horowitz and Meng research groups who exemplify the cooperative spirit of the Stanford Center for Integrated Systems. In particular, we thank Dr. Joe Ingino, Dr. Adrian Ong, Dr. Sha Rabii, Dr. Stefanos Sidiropoulos, Dwight Thompson, Katayoun Falakshahi, Jim Burnham, Sotirios Limotyrakis, Bendik Kleveland, Alvin Loke, Dr. Ken Yang, Gu-Yeon Wei, Birdy Amrutur, Dan Weinlader, Won Namgoong, Jeannie Ping-Lee, Greg Gorton, Syd Reader and Horng-Wen Lee. These friends and colleagues have provided a constant source of stimulating conversation that has no doubt improved the present work in tangible ways.

One person who deserves our special gratitude is Ann Guerra. Her administrative competence, contagious enthusiasm and warm sense of humor are welcome components of life at CIS, and we are grateful to her for her assistance, often given under great time pressure, and for her positive attitude. We are happy to add our thanks to those of countless others that have gone before us in acknowledging her central role.

Outside of CIS, we would like to thank Al Jerng, Allen Lu and the laboratory of Professor Leonid Kazovsky for their valuable assistance with our earliest low-noise amplifier work. In particular, the members of the Kazovsky laboratory were very generous in allowing us to use their facilities at a time when the SMIrC group had no laboratory of its own.

In addition to those in the Stanford community, we wish to thank several people outside Stanford who contributed directly to this work. In particular, we thank: Norm Hendrickson of Vitesse for supplying high-frequency packages; Howard Swain for his teaching and for first alerting us to the issue of induced gate noise; Dan Dobberpuhl of DEC for providing the opportunity to do some experimental work with DEC's 0.35μm CMOS technology, and Mark Pierce, Dave Kruckmeyer and the other members of the Palo Alto Design Center for helping us with simulation and tapeout; Dr. Chris Hull of Rockwell Semiconductor Systems for partnering with us for the GPS receiver work and attending to any problems that arose at Rockwell during the course of the project, and Paramjit Singh of Rockwell for his invaluable assistance with technology issues; Ernie McReynolds of Tektronix for helping us to insert a much-needed induced gate noise model into the BSIM-III code and for his general help on simulation issues; and Pauline Prather of New Focus who bonded many chips for us, often on very short notice, and never once complained about the extra work.

Finally, Dr. Shaeffer would like to express his gratitude for the steadfast support of his wife, Deborah Shaeffer, who patiently endured the many hours of preparation that went into this book.

Introduction

Derek K. Shaeffer

Wireless communications research has experienced a remarkable renaissance in the last decade. The advent of cellular telephony has driven much of the recent research activity, but substantial efforts have also focused on other wireless applications, such as cordless telephones and, more recently, the Global Positioning System.

The primary goal of this book is to explore techniques for implementing wireless receivers in an inexpensive complementary metal-oxide-semiconductor (CMOS) technology. Although the techniques developed apply somewhat generally across many classes of receivers, the specific focus of this work is on the Global Positioning System (GPS). Because GPS provides a convenient vehicle for examining CMOS receivers, a brief overview of the GPS system and its implications for consumer electronics is in order.

The GPS system comprises 24 satellites in low earth orbit that continuously broadcast their position and local time [4]. Through satellite range measurements, a receiver can determine its absolute position and time to within about 100m anywhere on Earth, as long as four satellites are within view. The deployment of this satellite network was completed in 1994 and, as a result, consumer markets for GPS navigation capabilities are beginning to blossom. Examples include automotive or maritime navigation, intelligent hand-off algorithms in cellular telephony, and cellular emergency (911) services, to name a few.

Of particular interest in the context of this book are embedded GPS applications where a GPS receiver is just one component of a larger system. Widespread proliferation of embedded GPS capability will require receivers that are compact, cheap and low-power. For such goals, the benefits conveyed by integration are self-evident: minimization of the number of off-chip components (particularly the number of expensive passive filters), improved form factor, reduced cost and ease of design.

For further cost reduction, it is interesting to consider implementation in a CMOS technology. Due to the huge capital investment in CMOS, it is only natural to consider whether the technology's shortcomings can be mitigated, making it attractive in an arena that historically has been dominated by more expensive silicon bipolar and GaAs MESFET technologies.

Meeting the goal of receiver integration in an inferior technology requires innovation in architectures, circuits and device modeling. Collectively, the scope of these problems is broad, but a successful approach will bring clear benefits for consumer electronics. And so, these considerations motivate the present research into highly integrated CMOS GPS receivers that forms the subject of this book.

The following chapters delve into the problems of radio receiver design in detail. The ultimate goal is the design and implementation of a 115mW CMOS GPS receiver in a 0.5-μm CMOS process. The techniques developed along the way are, however, broadly applicable to other wireless systems.

Chapter 1 begins with an overview of radio receiver architectures by presenting fundamental concepts through the vehicle of historical examples. Then in Chapter 2, the subjects of noise, distortion and frequency planning are presented, with special attention paid to cascaded systems. In addition, a review of the current state of the art in CMOS receiver research establishes a context for the present work. In Chapter 3, the relevant technical details of the GPS system are presented along with a brief survey of common GPS receiver architectures. Then, applying the concepts developed in Chapter 1, we introduce a new architecture that takes advantage of details of the GPS signal spectrum to achieve a high level of integration.

Chapter 4 tackles the subject of CMOS low-noise amplifiers in great detail. This includes a survey of recent work and the development of a power-constrained noise figure optimization procedure for gaining the best performance for a stated power budget. Proceeding down the receiver chain, Chapter 5 discusses frequency mixers and focuses attention on the double-balanced CMOS voltage mixer that provides high linearity, low noise figure and extremely low power consumption. Chapter 6 follows with an investigation of active filters. Because the active filter is a dynamic range bottleneck in many receivers, this chapter focuses on how to design filter transconductor elements that maximize dynamic range with a given power consumption. In particular, we develop a figure of merit that permits a comparison of various transconductors, leading ultimately to a very power-efficient filter implementation.

To put these theoretical developments into practice, Chapter 7 presents the implementation of an experimental CMOS GPS receiver in a 0.5μm process. The experimental results demonstrate a high level of performance and integration that is comparable to or better than existing implementations in more expensive technologies, thereby confirming the value of the techniques presented in earlier chapters. Finally, Chapter 8 concludes with a summary and some suggestions for future work.

For readers who survive the first eight chapters, several appendices present expanded treatment of certain subjects. Appendix A explores the topic of noise correlations in amplitude-limited gaussian noise channels. Appendix B presents a noise figure analysis of the MOSFET device using the classical technique. Appendix C presents some experimental results on two low-noise amplifiers: a single-ended amplifier and a differential amplifier. Finally, Appendix D describes the measurement techniques used to gather the experimental data reported in Chapter 7.

THE DESIGN AND IMPLEMENTATION OF LOW-POWER CMOS RADIO RECEIVERS

Chapter 1

RADIO RECEIVER ARCHITECTURES

The advent of wireless communications at the turn of the 20th century marked the beginning of a technological era in which the nature of communications would be radically altered. The ability to transmit messages through the air would soon usher in radio and television broadcasting and wireless techniques would later find application in many of the mundane tasks of everyday life. Today, the widespread use of wireless technology conveys many benefits that are easily taken for granted. From cellular phones to walkie-talkies; from broadcast television to garage door openers; from aircraft radar to hand-held GPS navigation systems, radio technology pervades modern life.

At the forefront of emerging radio applications lies modern research on the integrated radio receiver. The goal of miniaturization made possible by integrated circuit technologies holds the promise of portable, cheap and robust radio systems, as exemplified by the advent of cellular telephony in the mid-1980's. As miniaturization continues, embedded radio applications become possible where the features of multiple wireless systems can be brought to bear on a particular problem. One example is the use of a GPS receiver in a cellular telephone to permit the expedient dispatch of emergency service personnel to the caller's exact location.

The design of integrated radio receivers entails a number of important considerations. To provide a background for the discussion of such matters, this chapter explores the important features of modern radio receivers by presenting them in the context of their historical development.

1. THE RADIO SPECTRUM

The goal of any radio receiver is to extract and detect selectively a desired signal from the electromagnetic spectrum. This "selectivity" in the presence of a plethora of interfering signals and noise is the fundamental attribute that drives many of the tradeoffs inherent in radio design. Radio receivers must often be able to detect signal powers as small as a femtowatt while rejecting a multitude of other signals that may be *twelve orders of magnitude* larger! Because the electromagnetic spectrum is a scarce resource, interfering signals often lie very close to the desired one in frequency, thereby exacerbating the task of rejecting the unwanted signals.

The scarcity of the spectrum has grown steadily more important over time. Consider the situation at the turn of the century: when Guglielmo Marconi first succeeded in transmitting the letter "S" across the Atlantic Ocean on December 12th, 1901, there were virtually no radio transmitters in service, and thus the only interference to be contended with was atmospheric noise. The transmitter of choice was the spark-gap, which was hardly a spectrally-efficient technique. On the receiving end, a simple "coherer" – a glass tube filled with oxidized metallic filings – served to detect the electromagnetic pulses generated by the spark [5]. This detection technique was as unselective as the transmission technique was spectrally wasteful. A spectacular demonstration of the unselective nature of this type of radio system occurred during the 1901 Americas Cup yacht race when several independent parties tried to broadcast up-to-the-minute race coverage to shore using spark-gap transmitters. Needless to say, the transmitted information was lost in a cacophony of interference from the various transmitters so that no one was able to receive intelligible signals [6].

We will use this failure of an early radio system as the starting point for a history of radio receiver development. For as the number of permanent transmitting stations grew exponentially, from a scant 100 stations in the U.S. in 1905 to over 1100 stations only ten years later [7], the scarcity of spectrum and the accompanying drive to higher and higher frequencies (a drive that continues to this day) stimulated the development of radio receiver architectures that were increasingly sensitive and selective. With the advent of TV and radio broadcasting in the 1920s and 30s, the demand for radio technology grew beyond the ranks of the military and the hobbyists to the all-powerful consumer. The economic incentives for satisfying this demand added fuel to the fire, as evidenced by the rapid pace of technological progress, a renewed interest in "short-wave" radio [8], and a marked increase in patent litigation [7]. The technologies developed along the way in part to meet the increasing demands of radio reception – such as the vacuum tube (or audion, as it was originally called), the piezo-electric resonator, and later on the transistor and the integrated circuit – tell the story of electronics in general, not just of radio. Indeed, one of the first consumer products produced at the birth of the transistor age was a portable AM radio [9].

Today, the electromagnetic spectrum is crowded with literally millions of radio signals. Frequency use extends from about 3kHz up to 300GHz, or

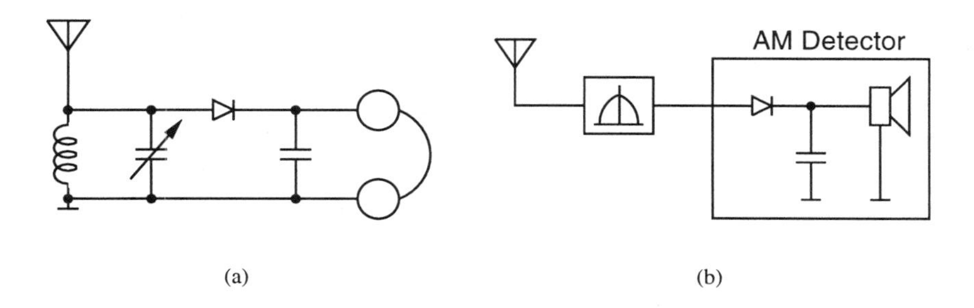

Figure 1.1. The crystal detector. (a) Schematic. (b) System diagram.

eight orders of magnitude in frequency. Of course, much of the research on integrated receivers today lies near the upper end of that frequency range. To understand how to design robust receivers in such a hostile environment, we will now consider some of the historical developments that have led to the radio architectures used today.

2. CLASSICAL RECEIVER ARCHITECTURES

The design of wireless receivers is a complex, multi-faceted subject that has a fascinating history. In this section, we will explore many of the fundamental issues that arise in receiver design through the vehicle of historical examples. These early receiver architectures illustrate an increasing level of sophistication in response to the need for improved selectivity at ever-greater frequencies. By considering their salient features, we lay the groundwork for a more formal treatment of the fundamental issues in the next chapter.

2.1 CRYSTAL DETECTORS

One of the earliest radio receivers is the crystal detector, shown in Figure 1.1. It is hard to imagine a more simple radio than this one. The received signal from the antenna is bandpass filtered and immediately rectified by a simple diode. If a sufficiently strong amplitude modulated radio signal is received, the rectified signal will possess an audio frequency component that can be heard directly on a pair of high-impedance headphones. The desired radio channel can be selected via a variable capacitor (or condenser, according to the terminology of the day). Remarkably, this radio does not require a battery; the received signal energy drives the headphones directly without amplification.

In the early 1900's, receivers of this type typically used diodes made of carborundum (silicon carbide) or galena (lead sulfide). Later on, with the advent of the vacuum tube, the "Fleming valve" or vacuum tube diode was sometimes substituted for the rectifying "crystal". Although exceedingly simple, the detector circuit used in this design was used in many of the more sophisticated radios that followed.

Though its simplicity is appealing, the crystal radio suffers from many important limitations that future architectures would seek to overcome. First, this receiver has very poor sensitivity. The rectified signal drives a pair of headphones directly. In addition, the received signal must be strong enough to periodically forward-bias the detector diode. These facts imposed a severe limitation on the transmission distance, or equivalently, on the transmitted power required for a given distance. At first, this burden was transferred to the transmitter side, where various techniques were developed to allow higher and higher transmit powers [7]. Later, the advent of the vacuum tube amplifier would permit the development of more sensitive receivers, thereby reducing the transmit power requirements.

Second, with only a simple bandpass filter, the crystal radio is not very selective. Accordingly, nearby radio channels may interfere with the desired channel. In addition, the use of spark gap transmitters, which persisted well into the 1920's, presented a pernicious source of interference due to the broadband nature of the spark signals. Although designs that followed would at first retain a similarly simple filtering approach, as frequencies increased the fractional bandwidth requirements for channel filtering would soon make the use of a single RF filter impractical.

With the advent of the vacuum tube triode (or audion, as its inventor liked to call it [10]), an early attempt to improve the sensitivity of the crystal radio took the form shown in Figure 1.2 [11]. This design was able to improve the audibility of received signals by about a factor of ten by using a galena crystal along with the audion. Signals from as far as 5,000 miles away could be detected with this simple technique.

It is interesting to note that this particular design was implemented at a time when the audion was very poorly understood and misconceptions abounded, many of which were perpetuated by its inventor. The first correct elucidation of the audion's behavior was given by Armstrong in 1914 [12], a mere ten days *after* the design in Figure 1.2 was presented. This serves to illustrate that even at the turn of the century, radio designers were working at the leading edge of electronics technology and were successful despite incomplete knowledge. This is often true today; indeed, certain aspects of this book demonstrate a similar situation, as will be shown later.

From the examples in this section, we see that two important limiting factors in radio design are the need for *sensitivity* and *selectivity*. The first of these factors was addressed to some extent with the advent of the heterodyne receiver.

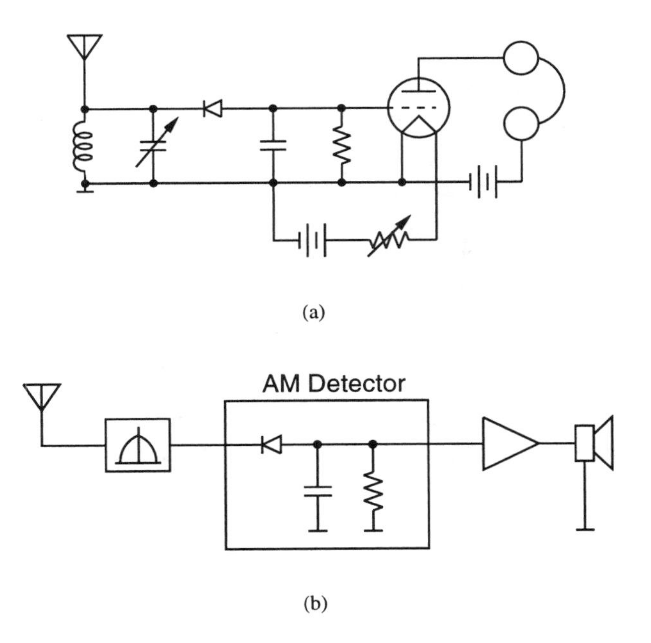

Figure 1.2. The crystal detector and audion amplifier. (a) Schematic. (b) System diagram.

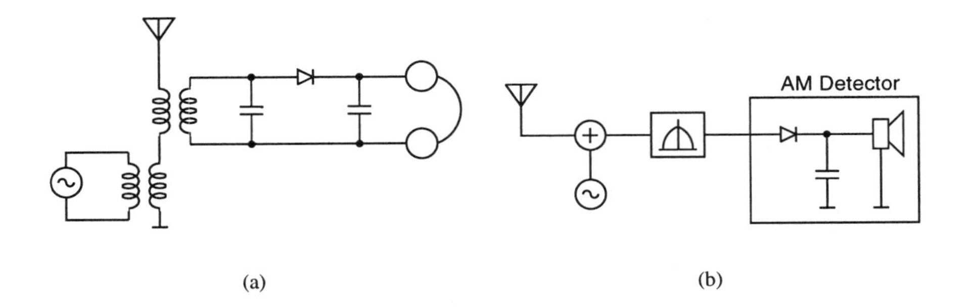

Figure 1.3. The heterodyne receiver. (a) Schematic. (b) System diagram.

2.2 HETERODYNE

The heterodyne receiver, shown in Figure 1.3, was first patented by Professor Reginald Fessenden in 1902 [13]. Initially used for wireless telegraphy, the receiver operates by summing a local oscillator signal with the received radio signal and rectifying the result. In the process of rectification, a "beat note" is produced in the headphones indicating the presence of the received radio signal. The operator of a receiver of this type could adjust the frequency of the local oscillator to select a comfortable pitch of the beat note.

The key advantage to the heterodyne receiver over a simple crystal detector was the increase in demodulation efficiency afforded by the use of the local oscillator. Even if the received signal was somewhat weak, the local oscillator could commutate the diode detector so that the beat note could be efficiently demodulated. So significant was this improvement that early treatises on the heterodyne's operation wrongly concluded that the apparatus increased the energy of the received signal, a startling claim that was later shown to violate conservation of energy [14].

One significant problem with the heterodyne receiver in Figure 1.3 is that the local oscillator is summed in series with the antenna, thereby rebroadcasting the local oscillator. However, this was not a serious problem when the heterodyne was first invented due to the fact that the received signal levels were typically very large (perhaps hundreds of millivolts). Later on, however, as the density of users and transmitters increased, and as transmitters began to operate on reduced power levels, isolation between the local oscillator and antenna became essential.

The most enduring feature of the heterodyne receiver is the use of frequency conversion under the control of a local oscillator. Although the initial purpose was to simply produce an audible tone for the detection of wireless telegraphy signals, the general concept of frequency conversion would later prove to be much more powerful when adapted for use in the superheterodyne receiver of Armstrong. Apparently, the general utility of this frequency conversion technique was unrecognized by the inventors.

In summary, the heterodyne receiver provided an increase in sensitivity by improving the efficiency of demodulation with the use of a local oscillator. It also introduced the concept of frequency conversion that would soon revolutionize the receiver art.

2.3 REGENERATIVE RECEIVER

Another innovation that sought to improve the sensitivity of radio receivers was the regenerative receiver, introduced by Armstrong in 1915 [15]. One version of this improved "audion" receiver is shown in Figure 1.4.

The regenerative receiver employed a single audion bulb for simultaneous use as a detector and amplifier. By placing a capacitor in series with the grid, the grid-to-filament circuit could be used as a simple detector with operation identical to a vacuum tube diode or "Fleming valve" as it was known at the time. However, unlike a Fleming valve or crystal detector, the audion would amplify the rectified grid signal in the plate circuit.

However, Armstrong was not happy with the improvement offered by this use of the audion alone. To increase the gain, he coupled the output of the

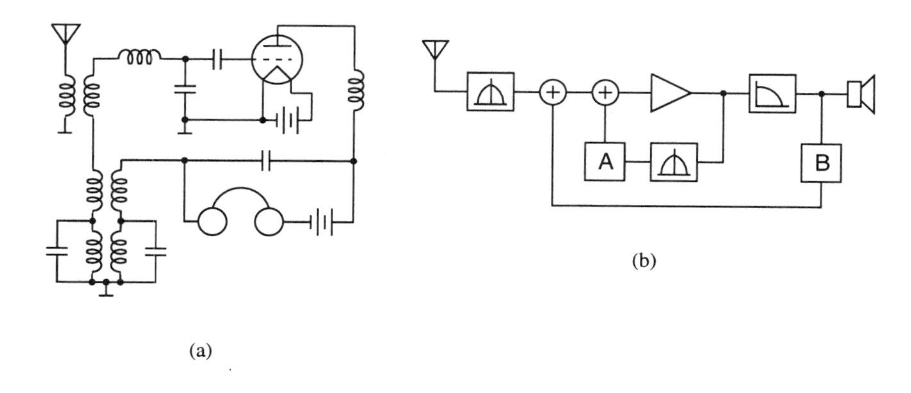

Figure 1.4. The regenerative audion receiver. (a) Schematic. (b) System diagram.

amplifier back to the input circuit with a radio-frequency transformer. In addition, he used a coil to increase the inductance of the input signal source. In modern terms, that inductance, when combined with the shunt capacitance in the grid circuit, forms an impedance transforming network called an L-match that increases the signal voltage available for amplification by the audion. A similar L-match appears in series with the plate circuit.

In another ingenious stroke, Armstrong also coupled the output of the amplifier back to its input with an audio-frequency transformer to pass the demodulated signal back through the amplifier once again. Thus, the single audion served as detector, RF amplifier and audio amplifier. To avoid interference with the RF feedback, the audio transformer was bypassed with capacitors at high frequencies.

In yet another variation on the basic regenerative receiver concept, Armstrong introduced a differential version that was able to reject static noise interference while retaining signal amplification. With this arrangement, Armstrong was able to receive signals at Columbia University from as far away as Germany and Hawaii. In a telling statement at the end of the paper, Armstrong attributes the success of his designs to "a proper understanding and interpretation of the key to the action of the audion".

Thus, we see that the important features illustrated by the regenerative receiver include the concepts of impedance transformation, gain boosting with positive feedback and the use of an active device for signal amplification and rectification; in modern terms, an active mixer. In addition, the selectivity of the system benefited from the use of multiple filters.

With the operating principle of the audion firmly established, and with the frequency conversion property of the heterodyne, the stage was set for the appearance of the basic radio architecture that is still used today in the vast majority of radio receivers: the superheterodyne receiver.

2.4 SUPERHETERODYNE

The superheterodyne receiver was invented by Armstrong in early 1918 [16] and the full technical details of the system were made public on December 3, 1919 [17]. A six-tube version of the receiver would later achieve wide commercial success as the first mass-produced AM radio. Despite the vast changes in electronics technologies since 1918, the superheterodyne architecture has endured and now forms the basis for almost all radio receivers made today.

In 1918, the detection of short-wavelength radio signals presented several challenges. The signal strength was generally much weaker than at longer wavelengths, making direct detection impractical and thus raising the need for more sensitive architectures. Direct amplification of short-wave signals was often impossible due to the limited frequency response of vacuum tubes available at the time. Finally, heterodyning of these signals required a very stable local oscillator that was difficult to implement (the superheterodyne preceded the advent of crystal resonators by a few years [18]).

Armstrong met these challenges in characteristically brilliant fashion with the superheterodyne architecture. A simplified version of his receiver with only a single amplifier stage is shown in Figure 1.5. The receiver employs a heterodyne front end that mixes the incoming radio signal with a local oscillator in a vacuum-tube detector to translate the RF signal to a pre-determined "intermediate frequency" where the signal can then be amplified and detected. By heterodyning to an intermediate frequency, the stability of the local oscillator becomes less important (though not irrelevant by any means, as we will see later on). Highly selective amplification and filtering can easily be obtained at the lower IF frequency so that detection of weak signals is made possible. In addition, through the use of multiple frequency conversions, the total required amplification can be distributed across several frequencies thereby aiding the stability of the amplifier stages and increasing the total possible amplification. Finally, by tuning the local oscillator to different frequencies, different RF signals could be selected for detection without having to re-tune the amplifier circuitry. This simplicity of adjustment opened the possibility of making a radio that could be used by unskilled operators. This feature would prove to be important in satisfying consumer demand for cheap and easy-to-use AM radios.

The concept of using multiple stages of frequency conversion to gain increased selectivity and extreme sensitivity is a powerful one that is widely used today. In addition, the concept of using a simple heterodyne detector to access radio signals that are beyond the frequency range of existing amplifier

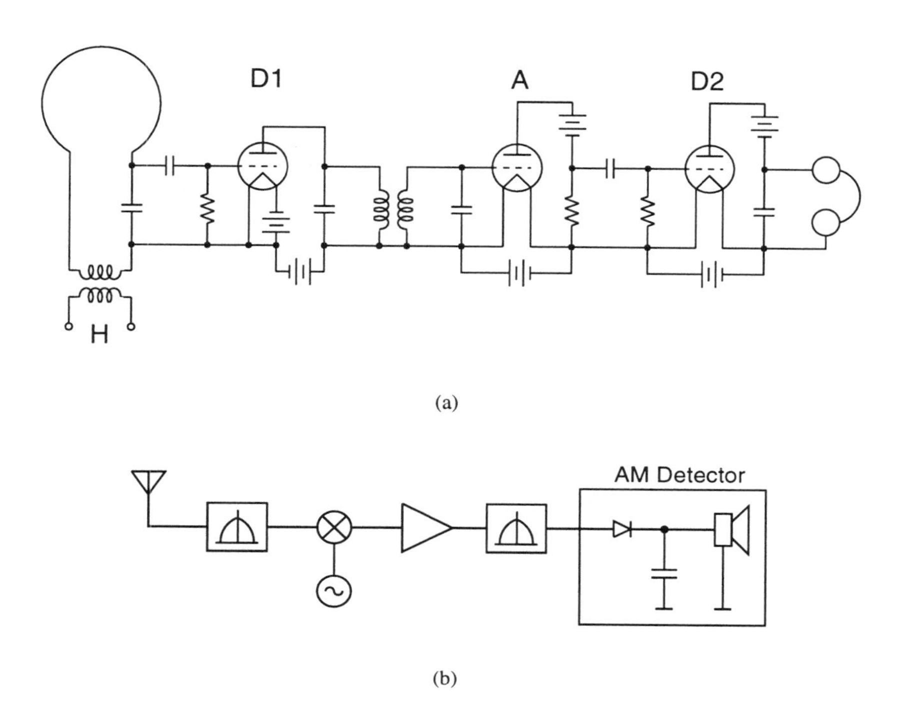

Figure 1.5. The superheterodyne receiver. (a) Schematic. (b) System diagram.

technology is still widely used in the millimeter wave frequency range from 30–300GHz [19].

2.5 SUPERREGENERATIVE RECEIVER

There is another receiver architecture due to Armstrong which, though not as enduring as the superheterodyne receiver, deserves honorable mention in the history of radio for its ingenuity. This architecture, known as the superregenerative architecture, employed a bizarre principle of amplification and detection in which a single vacuum tube was capable of producing power gains on the order of 100,000; a truly remarkable feat for a single triode tube [20]. A schematic of one incarnation of the superregenerative receiver is shown in Figure 1.6.

The first tube in this receiver is an RF oscillator whose oscillations are periodically quenched by a second oscillator, formed with a second tube, that runs at a lower frequency. At the end of each quench period, oscillations in the RF tube build up in response to initial conditions imposed by the incoming radio signal. Thus, after a fixed elapsed time imposed by the low frequency

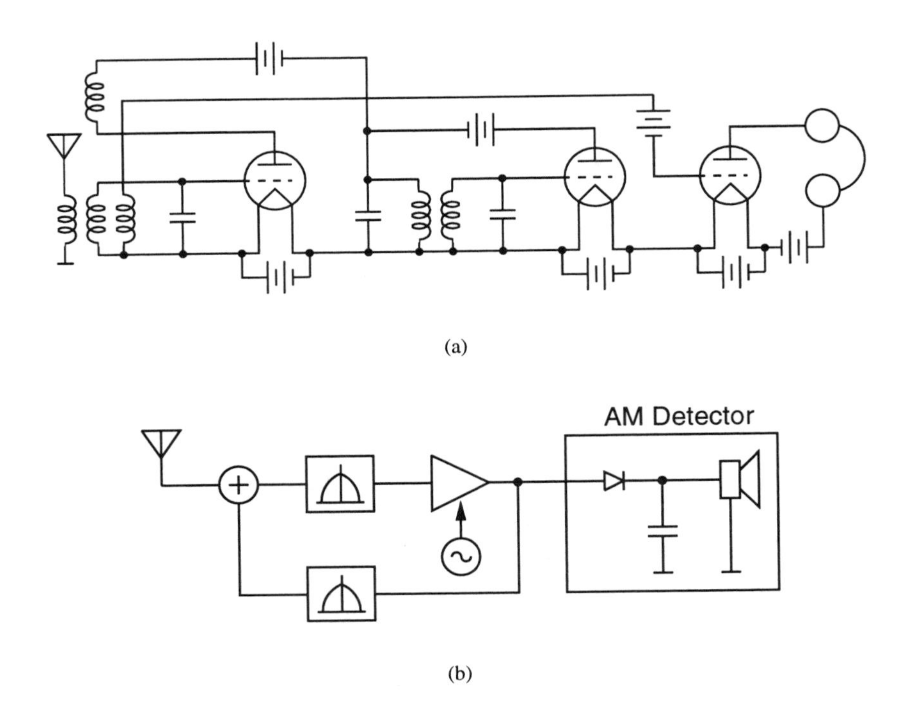

Figure 1.6. The superregenerative receiver. (a) Schematic. (b) System diagram.

oscillator, the RF oscillations build up to a level whose amplitude is proportional to the instantaneous amplitude of the received radio signal at the moment that oscillations began. The longer the time between quench periods, the greater the gain that can be achieved. In fact, the maximum gain depends exponentially on the relative frequencies of the two oscillators. The resulting output signal from the RF oscillator is a series of oscillation bursts whose amplitudes are proportional to the RF signal amplitude. The output can then be demodulated with a simple AM detector.

Remarkably, this receiver technique is essentially a sampled-data system. The radio signal amplitude is periodically sampled at the end of each quench period and the regenerative action of the RF oscillator amplifies these signals as the oscillation envelope grows exponentially. Because of the exponential growth, fabulous signal gains can be achieved. And, because the sampling rate is less than the carrier frequency but greater than the modulation bandwidth, the superregenerative receiver can be viewed as the first *sub-sampling* radio architecture.

2.6 AUTODYNE AND HOMODYNE

In the regenerative receiver, when the output is overcoupled to the input, the system oscillates, and this oscillation can be used to heterodyne the incoming RF signal. Such an arrangement was originally called an "autodyne", or automatic heterodyne, system [21]. In 1924, Colebrook observed that making the autodyne frequency equal to the RF frequency eliminated the need for an A.M. detector [22]. Thus, the homodyne receiver was born.

Unfortunately, for the homodyne to work effectively, the local oscillator must be precisely synchronized with the RF carrier. Any phase difference would lead to a reduction of the demodulated signal level. Recognizing the need for carrier synchronization in the homodyne receiver, a Frenchman named de Bellescize patented a version of the homodyne in 1930 that included carrier synchronization circuitry [23]. This receiver is shown in Figure 1.7.

The received RF signal is demodulated by a dual-grid tube in which a local oscillator is applied to the second grid. The demodulated output is lowpass filtered and the difference frequency (nominally D.C.) is applied to a control tube that tunes the oscillator tube to keep it synchronized with the received RF carrier. Hence, this technique is essentially the same as that used in modern phase-locked loops. Note that the selectivity of this receiver rests almost entirely on the audio lowpass filter.

The homodyne concept has been revived in recent years for application in paging receivers, which use a very simple FSK signaling technique. In applications requiring greater sensitivity and selectivity, the homodyne stands at a disadvantage due to its sensitivity to D.C. offsets and 1/f noise in the audio section. Also, because the local oscillator is tuned to the RF frequency, it can radiate back out the antenna and interfere with other receivers or reflect and be re-received and downconverted into a substantial, time-varying D.C. offset. These problems have prevented the widespread proliferation of the homodyne, although interest in the architecture has recently been revived [24].

2.7 SINGLE-SIDEBAND TRANSMISSION

With the growth of the radio art in the 1920's came the need to conserve the use of the spectrum. Economic factors motivated such conservation because a conservation of spectrum led in turn to increased capacity. One of the key developments that enabled a significant spectral savings was the advent of single-sideband transmission, which had its origin in multi-carrier wireline telephony [25]. The single-sideband transmission technique was apparently invented by John R. Carson in 1915 for use in the Bell System [26] and he filed patents for several inventions related to single-sideband transmission and reception in 1915 and 1916 [27] – [29]. For the present GPS work, the history of SSB transmission has direct relevance because the dual of the SSB transmitter

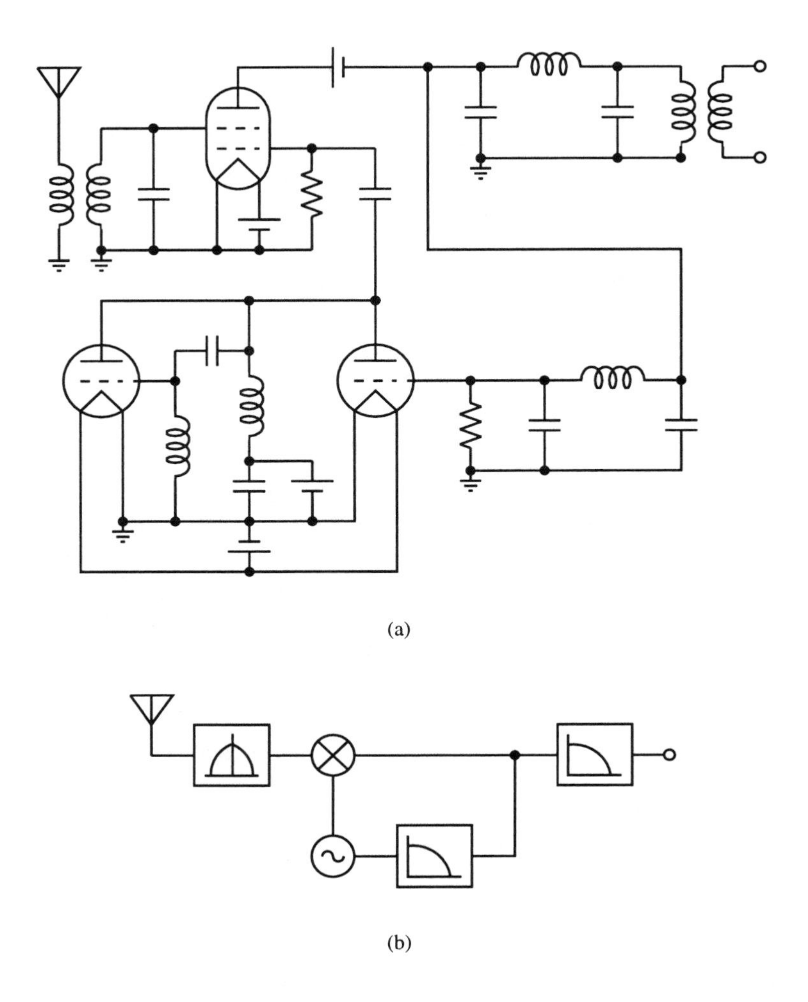

Figure 1.7. The homodyne receiver of de Bellescize. (a) Schematic. (b) System diagram.

is the image-reject receiver. The use of image rejection enables a high level of integration to be achieved, as will be shown in the following chapters.

A standard A.M. system produces a modulated signal comprising a carrier and two information-bearing sidebands: an upper sideband and a lower sideband. Because the carrier conveys no information, while the two sidebands convey redundant information, significant power and spectral savings can be had if the carrier is suppressed and one of the sidebands eliminated before transmission. The benefits of such an approach also include improved transmission distance because the transmit power required for the carrier and

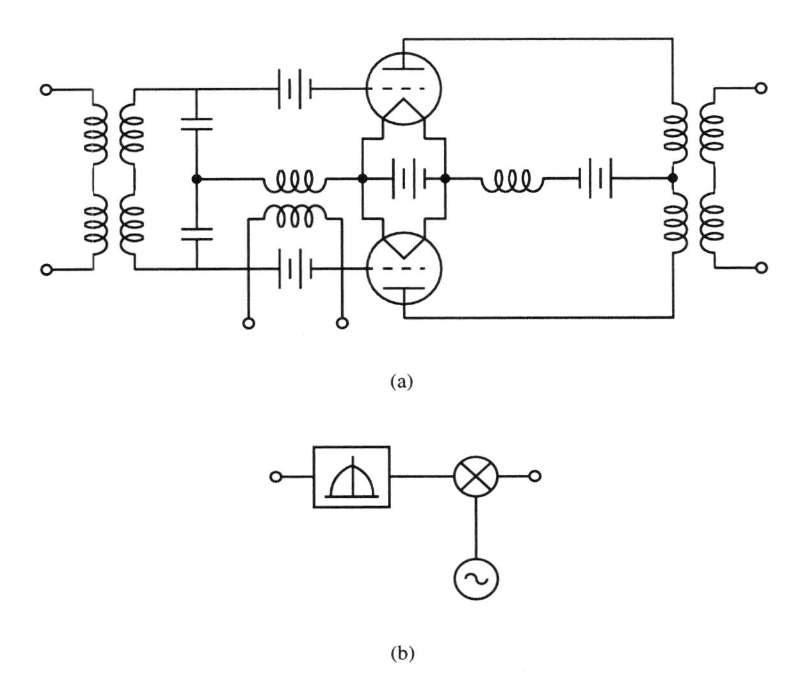

Figure 1.8. The single-balanced modulator. (a) Schematic. (b) System diagram.

rejected sideband may be reallocated for use in transmitting the remaining sideband [30].

One of the key inventions of Carson for SSB transmission that would find wide application in radio is the single-balanced modulator, shown in Figure 1.8. This modulator consists of two triode modulators with the modulating signal injected differentially and the carrier injected in a common-mode fashion. When the modulated output signal is then extracted differentially at the output of the two modulators, the carrier, being a common mode disturbance, is suppressed and does not appear in the output. Thus, the balanced modulator accomplishes the first task in SSB modulation: the rejection of the carrier.

In a typical SSB transmitter of the 1920's, the second task of sideband suppression would be handled by a simple filter that would pass the desired sideband and reject the undesired one. Due to the practical difficulties associated with filtering out one of the sidebands when the carrier frequency was high, practical transmitters used a sequence of frequency translations and filters to upconvert the modulated signal to a target frequency in stages while filtering out the unwanted sidebands generated at each step, thereby relaxing the required filter order [26]. This same principle can be applied to receivers as well to obtain greater selectivity. In fact, one might consider the superheterodyne receiver itself to be a dual of this type of SSB transmitter. Although this technique was very practical and widely used, it was expensive to implement due to the number of frequency translation steps involved.

So, two important contributions arising from the SSB transmission efforts of Carson and others are the balanced modulator and the use of multiple frequency translations to ease the filtering burden in selecting the desired sideband. Nonetheless, the expense of this early SSB technique led to the development of other approaches that did not rely on sharp filters. The first of these alternative approaches was the Hartley modulator.

2.8 HARTLEY MODULATOR

In 1925, Ralph V. L. Hartley invented a SSB modulator that replaced the more expensive filtering technique with a phase-shift technique that allowed direct cancellation of the unwanted sideband. His original system, taken from his 1928 patent [31], is shown in Figure 1.9.

The basic operating principle of the Hartley modulator is to produce two modulated signals: one with sidebands that are in phase with each other, and one with sidebands that are out of phase with each other. Then, by adding or subtracting the two modulated signals, one of the two sidebands can be reinforced while cancelling the other. The necessary phase shifts are most expediently introduced by two 90° phase shifters: one in the audio signal path and the other in the oscillator circuit.

In Hartley's original implementation, two bandpass L-C ladder filters (or "electric wave-filters" as they were then called [32][33][34]) provided a 90° phase shift between the two audio channels by the addition of an extra L-C section in one of the filters. The filter complexity was required to provide accurate quadrature over the whole audio band of interest. In contrast, the quadrature in the oscillator circuit was obtained with a simple R-C network.

The essential contribution of Hartley's modulator is the use of phase shifts to achieve cancellation of the undesired sideband. In modern terms, one could say that he introduced complex signal processing with his use of quadrature signal and local oscillator paths.

Unfortunately, a major drawback in the Hartley modulator is the need for filters that provide accurate, broadband quadrature while maintaining amplitude balance between the two audio channels [35]. This limitation was later removed by an innovative modification due to Donald K. Weaver, Jr.

2.9 WEAVER MODULATOR

In 1956, Weaver introduced another method for generating SSB modulation [36]. Interestingly, he referred to his method as a "third" method, in defer-

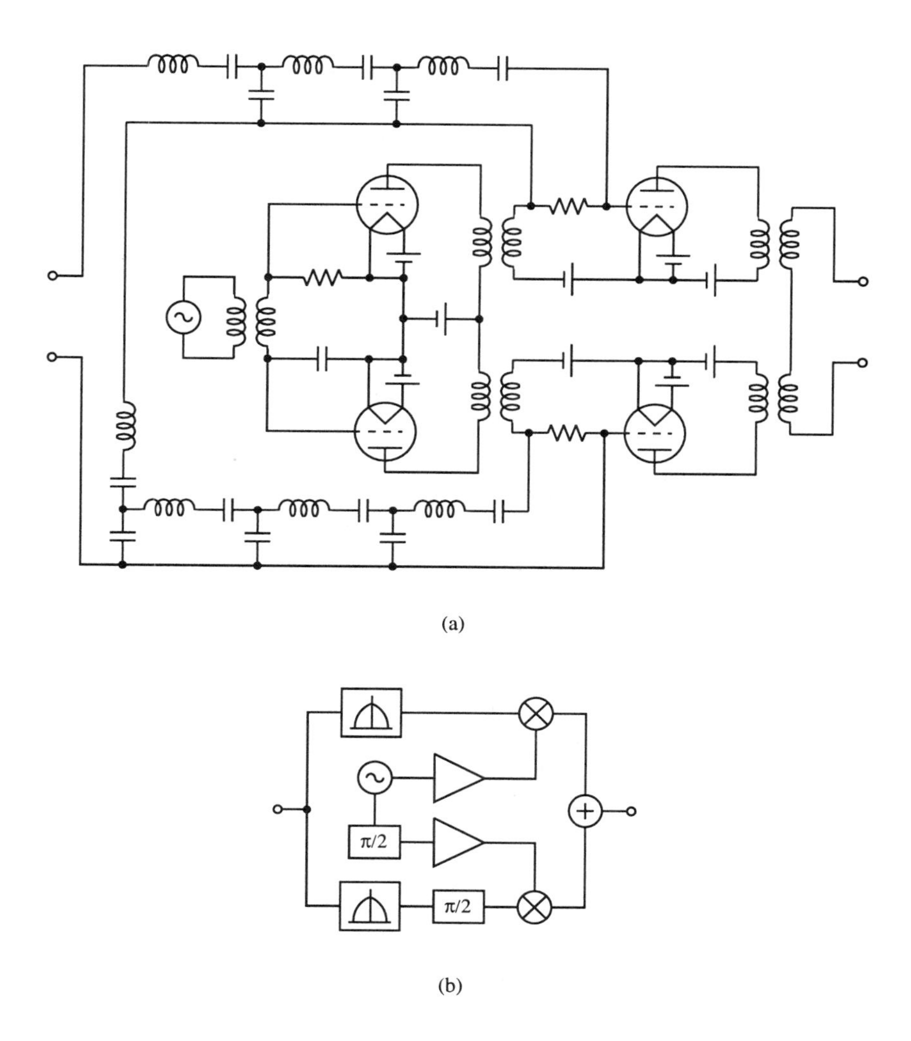

Figure 1.9. The Hartley SSB modulator. (a) Schematic. (b) System diagram.

ence to those introduced by Carson and Hartley, thereby implicitly ignoring other techniques that had been developed, such as one due to Kahn that used envelope elimination and restoration [37]. A schematic of Weaver's original implementation of his method appears in Figure 1.10.

In essence, the Weaver modulator replaces the broadband quadrature filter networks in the Hartley modulator with a quadrature frequency conversion. The generation of a quadrature first local oscillator is relatively simple because it operates at a fixed, single frequency. Hence, a simple R-C network suffices for

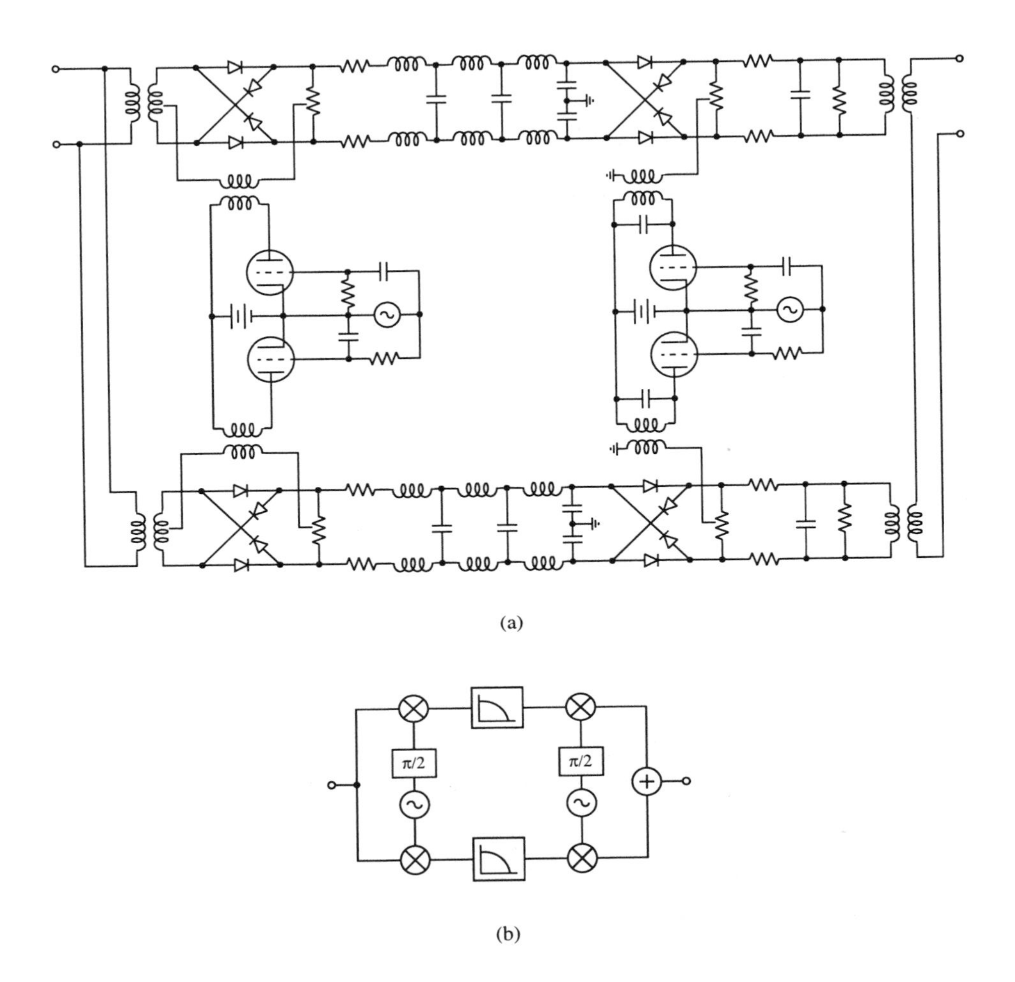

Figure 1.10. The Weaver SSB modulator. (a) Schematic. (b) System diagram.

quadrature generation. With the burden of quadrature generation now shifted to the local oscillators, the two signal paths can be more accurately matched for improved sideband suppression. Just as in the Hartley modulator, sharp filters are not required in this technique.

Although the original implementation employed a frequency *downconversion* from the audio band to DC, the principle of operation still holds for other choices of IF frequency. Indeed, the same architecture can be used as an SSB modulator or as an SSB receiver by reversing the sequence of frequency translation steps.

The Weaver modulator has become the most widely-used architecture for SSB receivers today. For a number of reasons that will be addressed in the next chapter, the Weaver receiver is the architecture of choice for a highly-integrated GPS receiver.

3. SUMMARY

In this chapter, we have seen how the developments during the early years of radio by Armstrong, Carson, Hartley, Weaver and others paved the way for the modern radio receivers. Collectively, these pioneers introduced the important concepts of frequency conversion, electrical filtering, balanced modulation and complex modulation are widely used in radio receivers today. Although the specific circuit implementations have changed, the basic principles remain the same. In particular, the superheterodyne architecture introduced by Armstrong is the most widely used receiver architecture today.

In the following chapter, we turn our attention to a more formal treatment of the fundamental issues that are introduced by these techniques which must be addressed in any successful receiver design. In addition, we will examine certain special issues that arise specifically in the context of integrated radio receivers.

Chapter 2

FUNDAMENTALS OF RADIO RECEPTION

The goal of this chapter is to provide a formal review of the essential concepts of noise, distortion, cascaded systems and frequency conversion. The selection of a suitable receiver architecture for a given radio standard is aided by a strong foundation in these topics. Hence, this section provides the background material for understanding the architectural tradeoffs discussed in the next chapter. We begin with the important topic of noise.

1. NOISE IN RADIO RECEIVERS

The sensitivity of all radio systems is limited by the presence of electrical noise that arises as a result of random fluctuations in current flow. Electrical noise can take on several forms including 1/f noise, thermal noise and shot noise. In radio receivers, our primary concern is generally with thermal noise which forms the subject of this section.

Surprisingly, the nature of thermal noise fluctuations was not understood until 1928 when Johnson and Nyquist published back-to-back papers describing experimental measurements and a statistical theory of noise [38] [39]. For example, in 1914, Lee de Forest boasted that "there appears to be no lower limit of the sensitiveness to the Audion, no minimum of suddenly applied e.m.f., below which the received impulses fail to produce any response." [10]. Sadly, he was quite mistaken.

The work of Johnson and Nyquist showed that all resistances in thermal equilibrium produce an *available noise power* that is proportional to the absolute temperature and the measurement bandwidth. Thus,

$$P_{av} = kTB \tag{2.1}$$

where k is Boltzmann's constant. Nyquist, in particular, produced an elegant and simple derivation of this fundamental relationship from first principles.

Two observations are in order. First, the fundamental quantity is the available noise power, which is the maximum power that can be delivered to a load impedance. This power has a value of 4×10^{-21} W/Hz at a temperature of T=290K. In the radio field, it is common to express signal powers in decibels, referenced to 1mW, which is typically denoted with the unit "dBm". Thus, the available noise power at T=290K is given by

$$P_{av} = 10 \log \left[\frac{4 \times 10^{-21} \mathrm{W/Hz}}{1 \times 10^{-3} \mathrm{W}} \right] = -174 \mathrm{dBm/Hz} \tag{2.2}$$

For a real resistance, the condition for maximum power transfer is that the load resistance be of equal value.[1] Because of this, the noise power can be attributed to an equivalent noise voltage in series with the resistor having a mean-squared amplitude of

$$\overline{v_n^2} = 4kTBR \tag{2.3}$$

or, equivalently, a noise current in parallel with the resistor having mean-squared amplitude

$$\overline{i_n^2} = \frac{4kTB}{R} \tag{2.4}$$

where R is the resistance value. Thus, although the available power is independent of resistance, the voltage or current is not. Secondly, the Nyquist relationship only holds for resistances that are in thermal equilibrium. This opens the possibility of producing by electronic means a real impedance that produces less noise power than a passive resistor because active electronics do not exist in a state of thermal equilibrium. This observation forms the basis of the art of low-noise amplification, in which an amplifier presents a specified input impedance that has an equivalent noise temperature associated with it that may be less than the ambient temperature.

In a radio receiver, the antenna also collects noise from the environment according to its power-directivity receiving pattern. Because the sky has a much lower noise temperature than the earth, the average noise temperature of an antenna will generally be less than the ambient temperature. To account for this difference, one can define an effective temperature for the antenna, T_a, that describes how much noise power it collects. Its available thermal noise power will then be given by

$$P_a = kT_aB. \tag{2.5}$$

[1]Note that for a complex impedance, maximum power transfer occurs when the load impedance is the complex conjugate of the source impedance.

In addition to noise collected by the antenna, the receiver electronics produce noise. To quantify the amount of noise thus introduced, North [40] introduced a quantity called noise figure, which is defined as

$$F \triangleq \frac{\textit{Total output noise}}{\textit{Total output noise due to the source}} \tag{2.6}$$

where the "source" is the antenna radiation resistance, under the (arbitrary) assumption that the antenna temperature, T_a, is 290K. This slightly chilly reference temperature was specifically proposed by Friis [41] because for this temperature $kT = 4 \times 10^{-21}$ W/Hz, a nice round number. Note that, for any passive network,

$$F = \frac{kTB}{kTBG_a} = \frac{1}{G_a} = L_a \tag{2.7}$$

where L_a is the available power loss of the network, defined as the available power at the input of the network divided by the available power at the output of the network.

A minor refinement of the language in (2.6) is necessary to avoid confusion in the case of mixer noise figures. The denominator in that case should read *total output noise due to the source that originates from the signal band of interest.* This distinction is necessary because mixers typically convert noise from multiple frequencies, as discussed in the section on frequency conversion.

With these definitions, the equivalent noise power at the antenna terminals of a radio receiver is given by

$$P_{eq} = kTB \left[\frac{T_a}{T} + (F - 1) \right]. \tag{2.8}$$

which reduces to simply $FkTB$ if $T_a = T$.

With the noise figure thus defined, it is a simple matter to specify the sensitivity of a radio receiver. If a specified minimum SNR is required for acceptable detection, the corresponding minimum detectable signal power is simply

$$P_{min} = \mathrm{SNR}_{min} \times kTB \left[\frac{T_a}{T} + (F - 1) \right] \approx \mathrm{SNR}_{min} \times FkTB \tag{2.9}$$

where B is equal to the effective noise bandwidth of the system. The approximation in (2.9) is only valid if $T_a \approx T$. For the nearly isotropic antennas used in most commercial GPS receivers, $T_a < T$ and the more exact expression should be used.

It is worth noting that the noise figure of any two-port network is determined by three quantities: the equivalent input voltage noise, the equivalent input current noise and the correlation coefficient relating the two noise sources. Because the correlation is generally complex, there are four parameters required to determine the noise performance of an arbitrary network [42]. Associated with these four parameters is a minimum noise figure that can be achieved and an optimum source impedance for achieving it. The reader is referred to [43] for the details of the classical technique, or to the Appendix for an example of a noise figure calculation for a simple MOSFET device.

2. SIGNAL DISTORTION AND DYNAMIC RANGE

If the thermal noise of a receiver sets the sensitivity, then the distortion introduced by the receiver sets the maximum signal level. The ratio between maximum and minimum signal levels defines the dynamic range of the receiver. This section explores the methods by which distortion is generated and formulates an expression for dynamic range that will prove useful later in the analysis of active filters.

It is common to assume that a distorting element has a transfer characteristic given by a simple power series

$$v_{out} = k_1 v_{in} + k_2 v_{in}^2 + k_3 v_{in}^3 \tag{2.10}$$

where k_1–k_3 are the gain, second- and third-order distortion coefficients, respectively. In such a case, if an input consisting of two closely-spaced sinusoidal components

$$v_{in} = A cos\,(\omega_1 t) + A cos\,(\omega_2 t) \tag{2.11}$$

is applied to the input, the output will contain several distortion products at frequencies $n\omega_1 \pm m\omega_2$, where $n + m$ is the order of the distortion product. Hence, in this case, $n + m \leq 3$. Furthermore, the amplitude of each product varies as A^{n+m}. So, second-order products vary in proportion to A^2 and third-order products in proportion to A^3 [44].

Figure 2.1 illustrates the behavior of the various intermodulation products with input amplitude. With the input and output amplitudes plotted on a log scale, the intermodulation product amplitudes follow straight line trajectories with slopes given by the order of the products. By extrapolating, intercept points can be found that serve as figures of merit for the linearity of the amplifier. These points can be referred to the input or output of the amplifier, as desired. Note that in a differential implementation, the second-order distortion is cancelled. Thus, in practice, second-order intercept points are typically much higher than third-order intercept points.

One aspect of third-order intermodulation distortion merits special attention. Among the third-order products are those that occur at $2\omega_1 - \omega_2$ and $2\omega_2 - \omega_1$. If $\omega_1 \approx \omega_2$, then these distortion products lie close to the fundamental

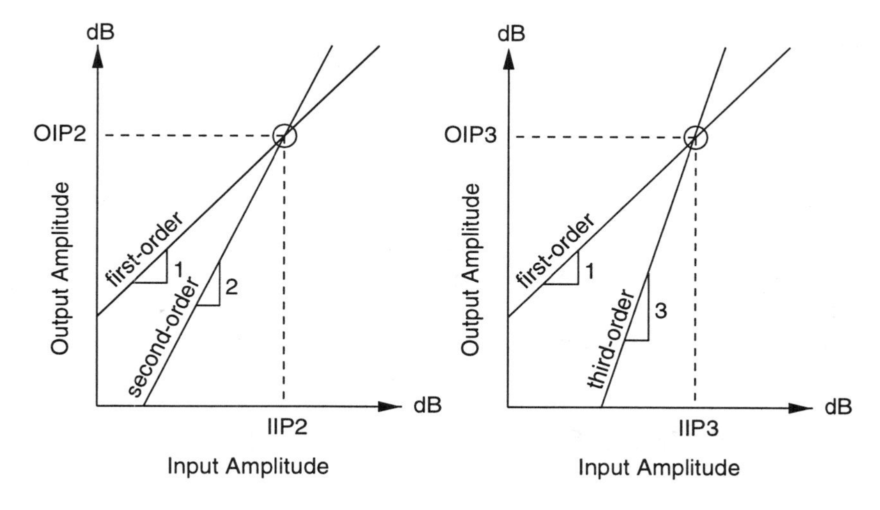

Figure 2.1. Illustration of intermodulation behavior.

tones in frequency and pass through any signal filters in the system virtually unattenuated. As a result, third-order nonlinearity represents a particular threat in radio systems.

If we assume that the distortion is dominated by third-order nonlinearity, we can formulate a useful expression for the spurious-free dynamic range of the amplifier. We define the peak SFDR as the difference between the maximum power level for which third-order intermodulation products lie below the noise floor and the minimum detectable signal power. The input-referred third-order distortion power level is given by

$$P_{\mathrm{IM3}} = \frac{P_s^3}{\mathrm{IIP3}^2}, \tag{2.12}$$

where P_s is the available source power and IIP3 is the available source power corresponding to the input-referred third-order intercept point. Setting this expression equal to $FkTB$ and solving for $P_s = P_{max}$, we obtain

$$P_s = P_{max} = \left[\mathrm{IIP3}^2 FkTB\right]^{1/3} \tag{2.13}$$

Hence, the peak SFDR is given by

$$\mathrm{SFDR}_{pk} = \frac{P_{max}}{P_{min}} = \frac{1}{\mathrm{SNR}_{min}} \left[\frac{\mathrm{IIP3}}{FkTB}\right]^{2/3} \tag{2.14}$$

An important caveat should be mentioned at this point. Although intercept points are useful figures-of-merit for amplifiers, they should be used with caution. In particular, practical amplifiers have gain and distortion curves that do not follow straight line trajectories when plotted on logarithmic axes. Thus, the intercept point loses its meaning unless the input or output power from which it is extrapolated is also specified. As a rule of thumb, the intercept points should be extrapolated from around the maximum anticipated operating power level of the amplifier in question.

As a final note, although third-order intermodulation distortion has special significance for radio receivers, there are some architectures that are particularly susceptible to second-order distortion. In particular, direct conversion receivers are sensitive to second-order distortion products that lie near DC because in such receivers the RF input is translated directly to DC itself [45].

3. FREQUENCY CONVERSION AND FREQUENCY PLANNING

In the previous subsections, we examined the topic of receiver sensitivity and dynamic range. The concepts presented there enable an evaluation of noise and linearity tradeoffs. In this section, we turn to the frequency domain to consider the topic of frequency conversion. In particular, we will look at the non-idealities introduced by practical frequency converters and the tradeoffs involved in the selection of intermediate frequencies and filtering strategies.

The modern term for the frequency converter is the mixer. All mixers operate on the principle that if two sinusoidal signals are multiplied together, the resulting product has sum and difference frequency components. Thus,

$$2cos\left(\omega_1 t\right) cos\left(\omega_2 t\right) = cos\left(\omega_1 t + \omega_2 t\right) + cos\left(\omega_1 t - \omega_2 t\right). \tag{2.15}$$

Note that if *one* of the cosines in the above expression is modulated in amplitude or frequency, the modulation is preserved in the output products. So, by multiplying an incoming radio signal with a local oscillator, one can translate the modulated signal to a different frequency for further processing.

There are two families of techniques for producing the desired multiplication: nonlinear techniques and time-varying techniques. In a nonlinear approach, the two sinusoidal signals are summed together and allowed to interact in a nonlinear device, such as a diode, vacuum tube or transistor. The resulting cross-modulation terms provide the desired frequency translation. However, the nonlinearity also produces signal distortion that is undesirable. In a time-varying approach, a variable gain block under the control of a local oscillator is used to produce direct modulation of the input signal. Because this technique is linear, cross-modulation between input signal frequencies and distortion of the input are avoided. As a result, the majority of modern mixers are of the time-varying variety. Nonlinear mixers find their primary use at very high frequencies and in low-cost applications, such as toy walkie-talkies and virtually every consumer AM radio. Figure 2.2 shows some examples of each type of

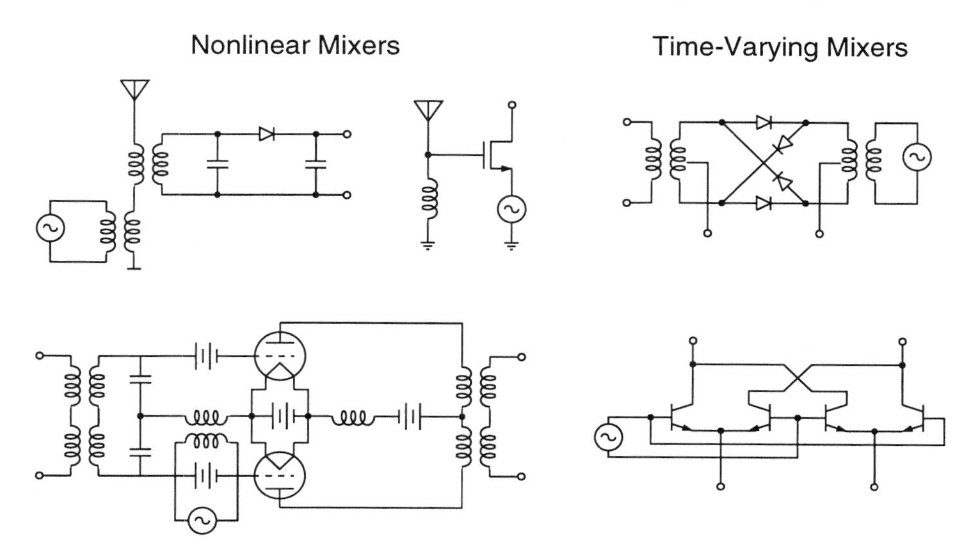

Figure 2.2. Various mixer topologies that fall into the nonlinear and time-varying categories.

mixer. Note that time-varying mixers may be of the voltage commutating type, such as the diode ring mixer, or of the current commutation type, such as the popular "Gilbert" mixer. [2]

All mixers are characterized in part by their *conversion gain*, which is the ratio of the desired output signal available power to the input signal available power. If we assume an ideal mixer of the switching variety that is internally lossless and has no bandwidth limitation and in which the instantaneous voltage gain from input to output alternates between one and minus-one at the local oscillator frequency, then the signal voltage at the IF port of the mixer is related to the signal voltage at the RF port by

$$v_{if}(t) = sgn\left(cos\left(\omega_{lo}t\right)\right) \times v_{rf}(t) = m(t)v_{rf}(t), \tag{2.16}$$

where the $sgn()$ function yields the sign of its argument. By performing a Fourier analysis of $m(t)$, we can easily determine that its fundamental component at ω_{lo} has an amplitude of $4/\pi$. Thus, by reference to equation (2.15), the voltage conversion gain is given by

$$G_c = \frac{2}{\pi} = -3.92\text{dB}. \tag{2.17}$$

[2] Technically speaking, the current-mode mixer shown in Figure 2.2 is not a Gilbert multiplier because it does not employ translinear principles but rather switches currents from one branch to the next under local oscillator control. A true Gilbert multiplier achieves a literal multiplication of two input signals using the translinear principle [46] [47].

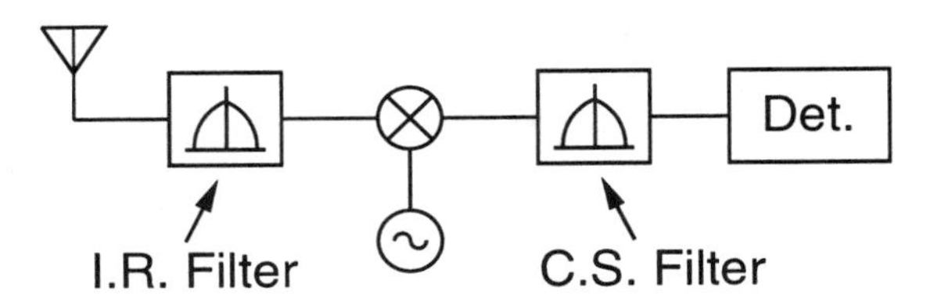

Figure 2.3. A simple receiver with image-reject filter and channel-select fitler.

In this example, the mixer is internally lossless and hence the voltage conversion gain is equivalent to the available power conversion gain.

Because the mixer produces both sum and difference frequencies at its output, there are two input frequencies that are translated with this conversion gain to the output at the *same* intermediate frequency, ω_{if}. These are $\omega_{lo} \pm \omega_{if}$. Typically, one of these frequencies is the desired RF signal while the other is commonly called the *image* frequency. Note that signals present at the image frequency can corrupt the intermediate frequency signal after mixing, making it desirable to reject the image before mixing.

One technique for doing so is shown in Figure 2.3, where the image frequency is removed with a simple bandpass filter. Because the image is separated from the desired frequency by $2\omega_{if}$, a *high* IF frequency relaxes the design of the image filter. On the other hand, a second filter at the IF frequency is typically used to select the desired signal and reject all other remaining signals. This channel-select filter is easier to implement if the IF frequency is *low*. Hence, the goals of selectivity and image rejection are in opposition and can be traded off by appropriately selecting the IF frequency. The essential aim of frequency planning is to select an IF frequency that adequately balances these competing requirements. One possible solution to this dilemma is to use *two* IF's, a high first IF to ease the image-reject filter design, and a lower second IF where channel selection can easily be done. This is a common approach in systems that have stringent specifications for image rejection and selectivity. Another possible solution is to eliminate the image-reject filter in favor of an image cancellation architecture, such as the Weaver SSB receiver. This is the approach taken in the present work for reasons that will be made clear in the next chapter.

In addition to frequency conversion, mixers also introduce extra noise that can degrade the noise figure of a receiver. This extra noise may originate due to losses internal to the mixer, or it may be directly down-converted from image bands at the mixer input. To understand this second source of noise, consider the case of an ideal mixer. If the mixer is internally lossless, then it contributes no noise of its own to the output and all of the output noise arises due to the source resistance. Secondly, this noise power is unattenuated by the mixer because the mixer only serves to periodically change the instantaneous sign of the white noise process without modifying its variance. Thus, it is tempting to conclude that the total output noise power is equal to the total output noise power due to the source, resulting in $F = 1$. We recall, however, that in the case of mixers we should restrict ourselves to considering the total output noise due to the source *that originates from the frequency band of interest.* If we assume that the frequency band of interest is centered about one of $\omega_{lo} \pm \omega_{if}$, then this component of the input noise spectrum is multiplied by the mixer conversion gain and appears attenuated at the output. Hence, we conclude that the true noise figure is actually

$$F_{SSB} = \frac{N_{out}}{N_{out,src}} = \frac{N_s}{G_c N_s} = \frac{1}{G_c} = 3.92dB \tag{2.18}$$

which is commonly called the *single-sideband* noise figure to indicate that the input frequency band of interest is only *one* of the two possible frequency bands that produce a response at ω_{if}.

In some systems the frequency bands of interest include those centered about *both* of $\omega_{lo} \pm \omega_{if}$. In this case, the output noise power originating from these two input frequency bands is twice as large (assuming that each contributes equally), and the resulting noise figure is

$$F_{DSB} = \frac{N_{out}}{N_{out,src}} = \frac{N_s}{2G_c N_s} = \frac{1}{2G_c} = F_{SSB} - 3\text{dB} = 0.92dB \tag{2.19}$$

which is commonly called the *double-sideband* noise figure, for reasons that should now be apparent. Note that the DSB noise figure is always less than the corresponding SSB noise figure, typically by about 3dB. In general, real mixers are not internally lossless and practical noise figures generally exceed these theoretical numbers for an ideal mixer.

There is yet another mechanism by which the mixing process can introduce noise to the system. As illustrated in Figure 2.4, a strong interfering signal (labeled "B" for blocker) can mix with local oscillator phase noise to produce noise that overlaps with the desired RF signal. The amount by which the noise floor increases as a result depends on the strength of the blocking signal. In particular, we can quantify the increase in noise power in a 1-Hz bandwidth due to the blocker

$$P_{n,b} = P_b \mathcal{L}\left\{f_{lo} - f_b\right\} = P_b \mathcal{L}\left\{\Delta f\right\} \tag{2.20}$$

In this expression, P_b is the blocker power at the mixer input and $\mathcal{L}\left\{\Delta f\right\}$ is the local oscillator phase noise power spectral density *relative to the carrier.* By setting (2.20) equal to FkT ($B = 1$Hz), we can determine the blocker power

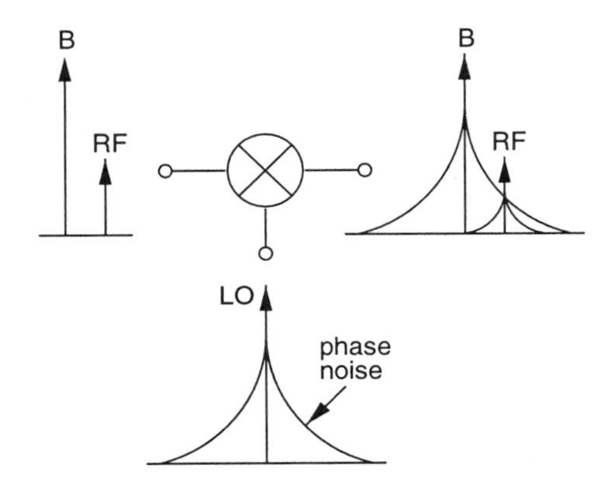

Figure 2.4. Illustration of the reciprocal mixing process.

that produces a 3-dB reduction in SNR, which is

$$P_b = \frac{FkT}{\mathcal{L}\{\Delta f\}}. \tag{2.21}$$

This expression can be used to determine the required phase noise specification for the local oscillator in order to achieve a given blocking performance. For example, a receiver with a 6dB noise figure and a 3-dB blocker level at 3MHz offset of -20dBm must have a local oscillator phase noise of better than -148dBc/Hz at 3MHz offset.

4. CASCADED SYSTEMS

When designing a radio receiver, it is often desirable to specify the performance of individual blocks (amplifiers, mixers, filters) separately to simplify the design task. The system performance is then determined by the cascade connection of these individual blocks, so it is important to understand the effects of cascading on figures-of-merit such as noise figure, linearity and dynamic range.

The noise figure of a cascade of signal blocks can easily be shown to be

$$F = F_1 + \frac{F_2 - 1}{G_{a1}} + \frac{F_3 - 1}{G_{a2}G_{a1}} + \ldots \tag{2.22}$$

where F_n is the noise figure of the $\mathrm{n^{th}}$ block *evaluated with respect to the driving impedance of the preceding block* and G_{an} is the available power gain of the $\mathrm{n^{th}}$ block. Note that available power gain is defined as the available output power divided by the available power from the source, where the available power is the power delivered to a matched impedance load. This definition is not the only definition for power gain [48], but equation (2.22), which is known as Friis's formula [41], is only correct when available power gain is used.

From (2.22), we can see that the first amplifier in a radio system contributes the most to the noise figure of the receiver; the contribution of each subsequent stage is reduced by the total available power gain preceding it. Thus, when designing for specific sensitivity, the greatest burden is borne by the first amplifier stage. For this reason, is important to have a low noise amplifier as close to the antenna as possible when maximum sensitivity is desired.

In a similar fashion, we can evaluate the linearity of a cascade of signal blocks. The production of intermodulation distortion in an amplifier cascade is somewhat more complicated, however, because the distortion products produced by each stage may have arbitrary phase relationships that make it difficult to precisely determine the cumulative distortion. However, with the simplifying (and somewhat optimistic) assumption that all distortion products add in *power* fashion, we arrive at the following expression [49]

$$\frac{1}{IIP3} \approx \frac{1}{IIP3_1} + \frac{G_{a1}}{IIP3_2} + \frac{G_{a1}G_{a2}}{IIP3_3} + \ldots \tag{2.23}$$

where $IIP3_n$ is the input-referred third-order intercept point of the nth stage, expressed in terms of the *available source power*, and G_{an} is the available power gain of the nth stage.

By re-expressing (2.23) in terms of *output* intercept points, we can demonstrate a certain symmetry between this expression and the one for cascaded noise figure. In terms of output quantities, (2.23) becomes

$$\frac{1}{OIP3} \approx \frac{1}{OIP3_n} + \frac{1}{G_{an}OIP3_{n-1}} + \frac{1}{G_{an}G_{an-1}OIP3_{n-2}} + \ldots \tag{2.24}$$

Now, at the receiver output, we must support a certain signal power for proper operation of the detector. This required output power plays a complementary role in linearity design to that of receiver sensitivity in noise figure design. In this case, as we work *backwards* from the detector, the contribution of each stage to the total $OIP3$ is reduced by the gain that *follows*. Thus, the last stage in the chain tends to contribute the most to the distortion and it is important to end the chain with an amplifier with high linearity.

Using equations (2.22) and (2.24), one can design a receiver that maximizes $IIP3$ and minimizes F, resulting in maximum dynamic range. As we've seen from these expressions, there is a natural tapering that occurs with early stages contributing more to the noise figure and later stages contributing more to the distortion, as illustrated pictorially in Figure 2.5. So, in general it is good for early stages to have good noise performance and later stages to have good distortion characteristics.

When selecting a gain plan for a receiver, there are three approaches that can be taken: design for minimum noise figure with acceptable linearity, design

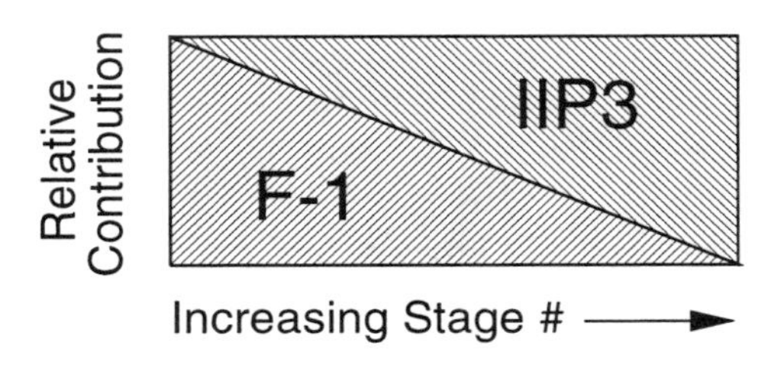

Figure 2.5. The relative contribution of successive stages to noise and distortion.

for maximum linearity with acceptable noise figure, or design for maximum dynamic range.

In the first case, one should taper the gains and noise figures of the stages so that the first stage dominates. However, it is important for linearity reasons not to be too greedy for gain. This approach is commonly used in receiver designs where sensitivity is paramount. In the second case, one should taper the gains and $OIP3$'s of the stages so that the last stage dominates. But one needs to be careful to use enough gain to meet the sensitivity requirements of the receiver. This approach may be particularly useful in applications where linearity is more important than noise figure.

In the last case, the condition for maximizing the dynamic range of a cascade of amplifiers can be determined analytically by considering a simple two-stage system. Suppose that we have two amplifiers with specified noise figures and input intercept points and that we wish to select the proper gain for the first amplifier to maximize the dynamic range of the cascade. It can be shown using (2.22) and (2.24) that the optimum first stage gain is given by

$$G_{a1} = \sqrt{\frac{(F_2 - 1)}{F_1} \frac{IIP3_2}{IIP3_1}} = \sqrt{\frac{CR_{21}}{CR_{12}}} \qquad (2.25)$$

where CR_{21} and CR_{12} are the *cross dynamic ranges* (similar to dynamic ranges) of the two amplifiers, as illustrated in Figure 2.6(a). By putting this expression in decibel form, the meaning becomes clear.

$$G_{a1} = \frac{1}{2}(CR_{21} - CR_{12}) \qquad \text{(in dB)} \qquad (2.26)$$

In words, the optimum gain is half the distance, in dB, between the two cross dynamic ranges. As shown in Figure 2.6(b), G_{a1} causes the output dynamic range of the first amplifier to be centered on a dB scale with the input dynamic range of the second amplifier. So, in a dynamic-range optimized system, each stage contributes noise and distortion equally.

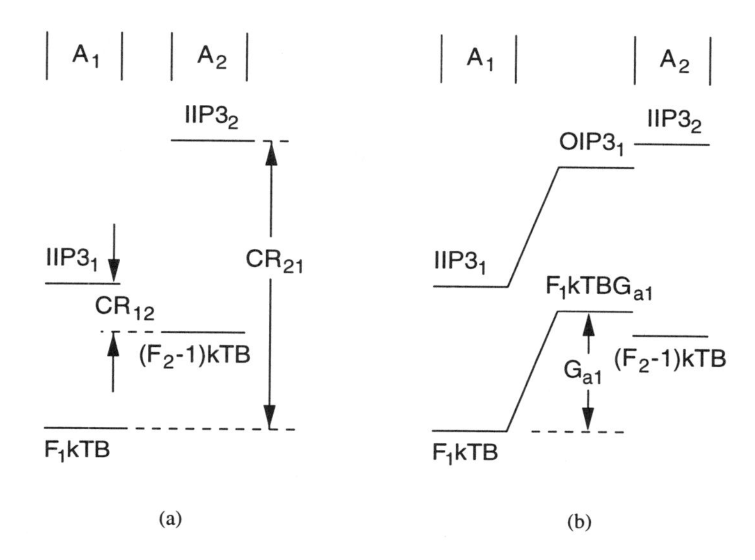

Figure 2.6. Optimizing dynamic range. (a) Two amplifiers with certain cross dynamic ranges. (b) Illustration of the optimum gain to maximize the dynamic range.

5. INTEGRATED RECEIVERS

Finally, we turn our attention to several issues that are particularly relevant to integrated receivers. The lack of high quality passive components and the presence of a common substrate profoundly influence integrated receiver design. Because the goal of this work is to achieve a high level of integration, the following sections briefly consider these factors to motivate the material in following chapters.

5.1 PASSIVE COMPONENTS AND THE FILTER PROBLEM

Most of the classical receiver architectures presented in the earlier sections rely heavily on high quality passive components in their implementations. In an integrated context, it is relatively easy to fabricate high quality fixed-value capacitors, but inductor options are rather limited. The two most viable sources of inductance for integrated radios are bondwires and spiral inductors. For bondwires, quality factors on the order of 50 are possible at 1GHz, but the achievable inductance values are limited to a few nanohenries and the absolute tolerance is also limited [50]. In contrast, spiral inductors offer relatively large inductance values and tolerances on the order of 5%, but quality factors are typically limited to less than 10 [51]. Recent work has demonstrated methods for improving the quality of spiral inductors using patterned ground shields [52], but spirals still have markedly inferior quality compared to off-chip passive components.

For these reasons, selective filters are difficult to achieve in integrated form. At frequencies near or above 1GHz, spirals and bondwires are very useful for reducing power consumption and providing limited selectivity, but at lower frequencies the required inductance values are out of the question. To integrate selective filters at frequencies below 100MHz, active filters are currently the only viable option. Unfortunately, active filters impose well-known dynamic range limitations that limit their utility [53]. In general, architectural choices that reduce the required fractional bandwidth of the active filters in the chain lead to improved dynamic range. Active filter issues are discussed in more detail in Chapter 6.

5.2 ISOLATION AND SUBSTRATE NOISE

Integrated radios, by their monolithic nature, have a common substrate that is shared by all of the circuit blocks on the chip. Because the substrate is not a perfect ground, isolation between blocks can be a particularly vexing issue [54]. Although substrate coupling is important in many mixed signal systems, it is especially difficult to mitigate in radio systems. Received signal levels may be on the order of a microvolt while digital signal amplitudes are typically more than a volt. To avoid desensitization of the front end due to digital noise coupling, it is therefore necessary to have isolation on the order of 120dB, which is very difficult to achieve.

To mitigate potential substrate coupling problems, a differential architecture is very helpful. The substrate presents a source of common-mode interference that may be partially rejected with the use of differential circuits. However, the goal of cointegrating the baseband DSP and the radio front end on the same chip remains elusive due to the extreme isolation requirements.

Another method by which isolation is compromised is through spiral inductor parasitics. Because spirals are relatively large structures (on the order of 300μm square), their capacitive parasitics to the substrate can be quite large. Fortunately, this source of coupling can be nearly eliminated through the use of patterned ground shields [52].

In addition to compromising isolation, the substrate introduces resistive losses that can reduce the quality factor of on-chip reactive components. This is particularly true of spiral inductors and varactors. Furthermore, the parasitic capacitances associated with bondpads and electro-static discharge (ESD) circuits couple into the substrate and the reflected losses can significantly impair the performance of low-noise amplifiers. So, it is important to shield critical bondpads from the substrate and carefully optimize ESD circuits to minimize their capacitive parasitics whenever possible to reduce these vexing sources of loss.

5.3 POWER, VOLTAGE AND CURRENT

In the previous sections, design criteria were presented that depend fundamentally on the *available power gains* of the individual blocks in a receiver signal chain. There appears to be widespread confusion in the recent literature about the applicability of power gain in non-50Ω environments. This section seeks to clarify some common misconceptions that seem to persist in the recent literature. Because of these misconceptions, it seems necessary to spend a few paragraphs understanding their source in the interest of clear communication of the experimental results of this book.

There is a fundamental reason for why power quantities are useful in radio design: the available noise power from any passive network at thermal equilibrium is kTB, independent of the resistance or reactance of the network. Note that, in particular, this available power is independent of the terminating impedance of the network. It is for this explicit reason that Friis, in his classic paper on noise figure, introduced available power gain in the definition of noise figure. To quote Friis, "it is the presence of such mismatch conditions in amplifier input circuits that makes it *desirable* to use the term available power" (emphasis added) [41].

It is important to recognize that impedance mismatch is a motivator of the use of power quantities, because it has often been argued that power gain is irrelevant in integrated radios because integrated radios do not use matched impedances. But as Friis clearly states, the opposite is in fact true: it is the presence of mismatches that motivates the use of power terminology! This is not to say that voltage and current quantities lack utility, but rather that power terminology often provides a natural and helpful viewpoint.

The widespread misunderstanding of this point stems, in part, from confusion about power gain itself. The problem is that there are, in fact, at least three definitions of power gain that are useful in different situations. These are available gain, operating gain, and transducer gain, which are respectively defined as [48]:

$$G_a = \frac{P_{\text{AVN}}}{P_{\text{AVS}}} = \frac{\text{power available from the network}}{\text{power available from the source}} \tag{2.27}$$

$$G_p = \frac{P_{\text{L}}}{P_{\text{IN}}} = \frac{\text{power delivered to the load}}{\text{power input to the network}} \tag{2.28}$$

$$G_t = \frac{P_{\text{L}}}{P_{\text{AVS}}} = \frac{\text{power delivered to the load}}{\text{power available from the source}} \tag{2.29}$$

Of these, confusion between operating gain and available gain seems to be the root of the problem. Note that the operating power gain, G_p, is defined with respect to the power input to the network. Hence, if the network presents an input impedance that is highly reactive (such as the gate of a MOSFET), then the operating power gain can become poorly defined. This argument is sometimes used in support of the notion that power gain is irrelevant in integrated CMOS amplifiers. However, the power gain used in the design equations of the previous sections is *available power gain*, which is perfectly well-defined, even when the operating gain is not. Hence, available power gain remains a useful quantity for designing amplifiers with reactive input impedances.

But, perhaps the most startling viewpoint of all is that power metrics – and in particular, power *units* – are only valid in matched, 50Ω systems. This viewpoint has been used by some authors to justify a modification of a standard unit of measure, the dBm, which is correctly defined as *the signal power, referenced to 1mW, expressed in dB*. Unfortunately, it is becoming widespread to use dBm to mean *the signal power dissipated in a* reference *50Ω resistor*. When used this way, dBm does not refer to the signal power at all, but rather the signal *voltage*, referenced to a hypothetical power level in a fictitious 50Ω resistance without regard for the actual signal impedance. Unfortunately, not all authors that employ this non-standard definition are explicit about their assumptions, which leads to unintentional confusion in the presentation of experimental results and tends to undermine the goal of clear scientific discourse. Other embarrassments that accompany the use of the corrupted unit may include an apparent violation of conservation of energy (e.g. amplifiers that deliver more output "power" than they consume from the power supply) and a complete invalidation of Friis's formula in cases where current noise is significant. Thus, we can see that a redefinition of standard units of measure leads to a plethora of unfortunate consequences.

So, in the interest of avoiding these problems, power gain will generally refer to *available gain*, unless stated otherwise, and the dBm unit will be used with its only true definition: the signal power, referenced to 1mW, and expressed in dB.

6. REVIEW OF RECENT CMOS RECEIVERS

In this section, we establish a context for the results to be presented in the following chapters by briefly reviewing some of the recent work on CMOS radio receivers. There are essentially three approaches to integrated CMOS receiver design that have recently been pursued: image rejection, sub-sampling and direct conversion. This section presents examples in each category from the recent literature.

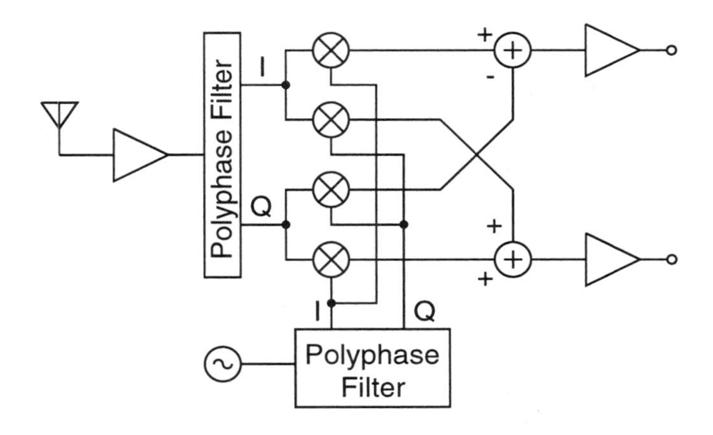

Figure 2.7. A low-IF image-reject architecture using polyphase filters.

Image reject receivers seek to eliminate multiple stages of frequency conversion and off-chip filtering in favor of image cancellation and a single stage of on-chip filtering. One example of this approach is illustrated in Figure 2.7 [55]. In this architecture, the 900MHz RF signal is amplified and filtered with a polyphase filter that produces a complex signal at its output. This signal is then converted to a low intermediate frequency of 250kHz with a set of complex mixers that essentially multiply the RF signal by $e^{j\omega t}$. Quadrature local oscillators are generated using another polyphase filter to achieve a $0.3°$ phase accuracy.

There are two drawbacks of this architecture that are worth mentioning. First, the use of a polyphase filter in the signal path contributes to a high noise figure due to the signal loss in the filter. Second, the image rejection is limited to about 46dB, even after correcting for amplitude errors in the two signal paths. The achievable image rejection is fundamentally limited by on-chip component matching and the phase accuracy of the quadrature-generating polyphase filters.

A second approach, shown in Figure 2.8, replaces the polyphase filter in the signal path with a pair of quadrature mixers and a second local oscillator [56]. An interesting feature of this 1.9GHz architecture is that no filtering is performed at the first IF. This "wide-band IF" is used to make the receiver amenable for use in a multi-mode radio. The use of a second set of mixers permits the first local oscillator to be somewhat removed from the desired RF frequency so that LO leakage back to the antenna does not present a severe threat. Channel selection is performed at baseband by a discrete-time switched-capacitor lowpass filter just prior to A/D conversion. This implementation achieves an image rejection of about 55dB.

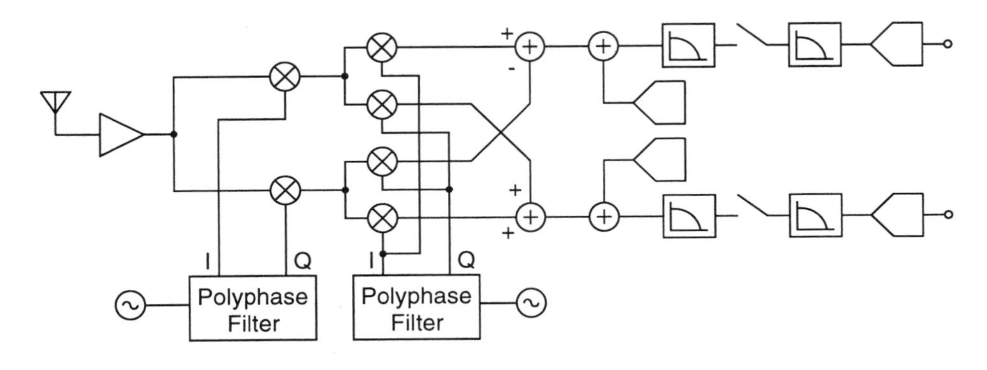

Figure 2.8. A low-IF image-reject architecture with a wide-band IF.

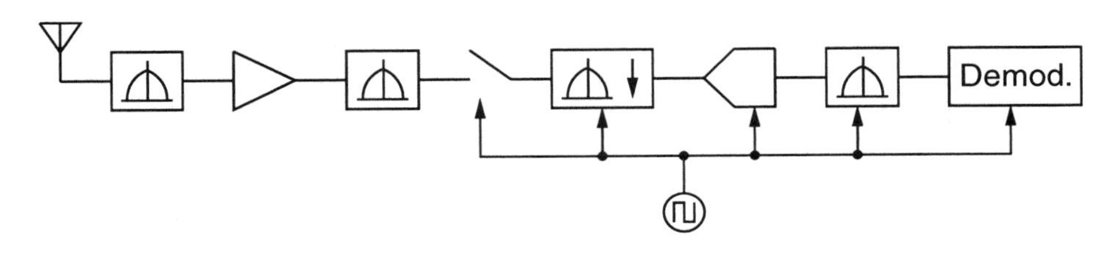

Figure 2.9. A sub-sampling receiver with discrete-time filtering.

A completely different approach is illustrated in Figure 2.9, which shows a sub-sampling receiver architecture in which most of the filtering is performed by discrete-time switched-capacitor filters [57]. The 910MHz RF input signal is amplified, filtered and directly sampled before any frequency conversion is performed. By sub-sampling the signal at 78Ms/s, an aliased image of the RF signal is produced at 26MHz, making the frequency conversion implicit in the sampling process. A sequence of down-sampling discrete-time filters follows, leading ultimately to an A/D converter.

The primary difficulty in this approach is the use of sub-sampling which causes broadband kT/C noise to alias into the signal band resulting in a very high noise figure of 47dB. The effect of clock jitter on the receiver noise floor is also exacerbated as a result of the sub-sampling.

Finally, in a revival of a technique that has found some success in paging receivers [24], the receiver shown in Figure 2.10 employs a homodyne, or direct-conversion technique in which the first (and only) local oscillator is tuned exactly to the RF carrier frequency [45]. The advantage of this technique is that there is no image frequency at all, making the architecture very amenable to integration. However, a significant problem is that low-frequency 1/f noise and offsets in the baseband section can overload the receiver, thus desensitizing

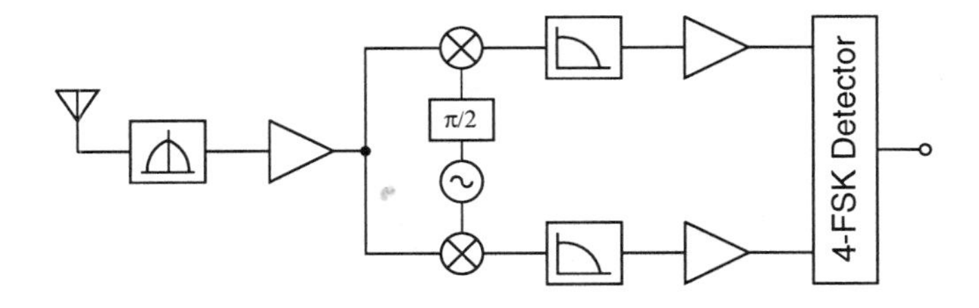

Figure 2.10. A direct-conversion receiver.

Table 2.1. CMOS Receiver Summary

Reference	Freq. GHz	NF dB	IIP3 dBm	I.R. dB	Gain dB	Pwr/Vdd mW/V	Area mm^2	Tech.
Crols [55]	0.9	24	28	46	9	500/5	6	$0.7\mu m$
Rudell [56]	1.9	14	-7	55	78	198/3.3	15	$0.6\mu m$
Shen [57]	0.9	47	-16	32	36	90/3.3	3.6	$0.6\mu m$
Rofougaran [45]	0.9	8.6	-8.3	N/A	140	360/3.3	77	$1.0\mu m$

it. One troublesome source of such offsets is LO self-mixing where the local oscillator leaks to the RF port of the mixer and then self-converts into a DC offset at the mixer output. To prevent this and other offsets from overloading the baseband amplifiers, offset cancellation must be used. The offset cancellation loop must have a very low bandwidth to avoid interference with the desired signal. This is achieved with a large off-chip $140\mu F$ capacitor.

In this direct-conversion receiver, the signaling technique is 4-FSK, which has the benefit that the modulation contains no energy at D.C. so that the offset cancellation loop does not corrupt the desired signal. Another interesting feature of this receiver implementation is the use of on-chip suspended spiral inductors with pits etched in the silicon beneath them. This etching reduces the spiral parasitics in return for a substantial area penalty.

The performance of the CMOS receivers presented in this section is summarized in Table 2.1.

7. SUMMARY

Building on the historical foundation of Chapter 1, this chapter has explored the mathematical fundamentals of radio reception, including such topics as noise figure, linearity, dynamic range and frequency planning. In addition, a review of some recent CMOS radio receivers provides a context for the present work.

Based on the prior art in CMOS radios, the low-IF architecture is the most appealing approach for channels with radio signals whose modulation contains significant energy near DC. Such is the case in the GPS system. In addition, there are other features of the GPS system that lend themselves quite naturally to a low-IF architectural approach. These motivating factors, and the specific architecture that results from them, are the primary considerations of the next chapter.

Chapter 3

A GLOBAL POSITIONING SYSTEM RECEIVER ARCHITECTURE

Successful radio designs begin with good architectural choices. Unfortunately, there is no radio architecture panacea. Rather, it is essential to select the approach best suited for the task at hand.

In this chapter, we turn our attention to selecting the GPS radio architecture that will permit the maximum level of integration while minimizing power consumption. We begin with the details of the GPS system itself. As will be shown, the GPS system possesses certain unique features that make it particularly well suited for integration.

1. THE GLOBAL POSITIONING SYSTEM

To motivate the architectural choices described in this chapter, it is important to consider some details of the received GPS signal spectrum. The GPS system uses a direct-sequence spread spectrum technique for broadcasting navigation signals. In such an approach, the navigation data signal is multiplied by a pseudo-random bit sequence (PRBS) code that runs at a much higher rate than the navigation symbol rate. This higher rate is commonly referred to as the "chip" rate of the code. The PRBS codes used in the GPS system are Gold codes that have two possible values (± 1) at any give time. Thus, when a code is multiplied by itself, the result is a constant value; however, when two different codes multiply each other, the result is another PRBS sequence. This property can be used to separate overlapping received signals from multiple satellites into distinct data paths for navigation processing. In principle, by multiplying the received signal by a particular satellite's PRBS code, the receiver can recover data from that satellite alone while signals from other satellites pass through with the appearance of pseudo-random noise. Hence, with a unique PRBS code assigned to each satellite, all satellites can broadcast at the same frequency without substantially interfering with each other.

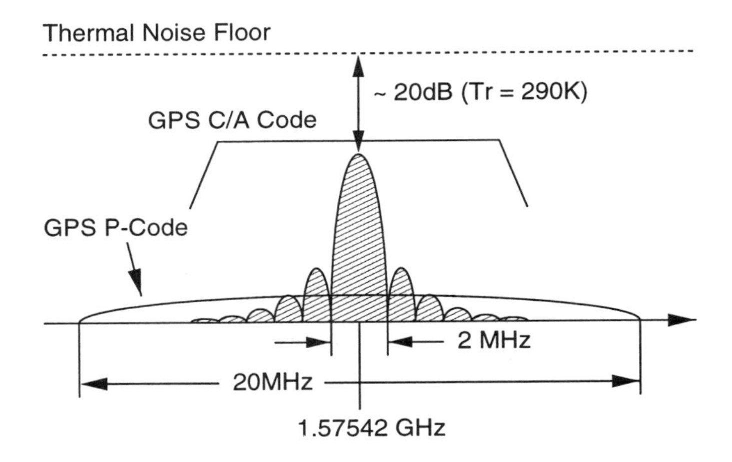

Figure 3.1. The GPS L1 band signal spectrum.

The GPS satellites broadcast navigation signals in two bands: the L1 band, which is centered at 1.57542GHz, and the L2 band, centered at 1.2276GHz. Each satellite broadcasts two different direct-sequence spread-spectrum signals. These are known as the P code (or precision code) and the C/A code (or coarse acquisition code). The P code is broadcast in *both* frequency bands, while the C/A code is broadcast only in the L1 band. Note that the center frequencies of the L1 and L2 bands are both integer multiples of 10.23MHz, which is the chip rate of the P code signal. In contrast, the C/A code uses a lower chip rate of 1.023MHz. The P code is intended for military use and is much more difficult to detect, in part because it uses a spreading code that only repeats at 1-week intervals. In addition, the P code is encrypted to restrict its use to authorized (military) users. For this reason, the C/A code is of primary interest in commercial applications.

Figure 3.1 illustrates the spectrum of the GPS L1 band. In this figure, we see that the C/A code and the P code occupy the same 20-MHz spectrum allocation, but their main lobes have different bandwidths due to the different code chip rates. In particular, the C/A code has a main lobe width of 2MHz while the P code has a width of 20MHz. The outlying lobes of the P code are truncated by appropriate filtering so that the entire GPS broadcast fits neatly within the 20-MHz allocation.

The immunity to interference that is gained when using the spread-spectrum technique is related to the ratio of the chip rate to the symbol rate. This ratio, called the *processing gain*, gives an indication of the improvement in SNR that occurs when a signal is "de-spread". For the GPS C/A code, the symbol rate is a mere 50Hz. Thus, the processing gain is given by

$$G_p = 10log\left(\frac{f_c}{f_b}\right) = 43\text{dB} \tag{3.1}$$

where f_b is the symbol rate of the C/A code, and f_c is the chip rate.

The received signal power is typically –130dBm at the antenna of a GPS receiver. If we assume that we are primarily interested in the 2-MHz main lobe of the C/A code, the noise power in this 2-MHz bandwidth is simply given by $kTB \approx -111\text{dBm}$ ($T = 290\text{K}$). Hence, the received SNR at the antenna is about –19dB. Once the signal from a given satellite is correlated with its PRBS code, the bandwidth is reduced to only 100Hz. Thus, the postcorrelation SNR improves by the processing gain of the system. So, with an antenna temperature of 290K and an otherwise *noiseless* receiver, the post-correlation SNR would be about 24dB.

2. TYPICAL GPS RECEIVER ARCHITECTURES

With the GPS signal spectrum in mind, consider two architectures that are widely used in commercial GPS receivers today. These are illustrated in Figure 3.2.

The first, and more widespread, is the dual conversion architecture. In this approach, the GPS L1 band is translated to a moderate intermediate frequency (IF) of approximately 100-200MHz where it is filtered off-chip before a second down-conversion to a lower IF of around 1-10MHz. There, the signal is filtered a second time before being amplified to a detectable level.

The second approach is the single conversion architecture. As the name implies, only a single IF is used, generally with off-chip filtering. The IF is directly sampled and then converted to baseband in a subsequent digital step. An alternative approach sub-samples the IF directly to baseband.

Both architectures have several attributes in common. First, an off-chip LNA or active antenna is generally postulated, permitting remote placement of the antenna from the receiver itself. It is also common to have several off-chip filters, including IF filters, phase-locked loop (PLL) loop filters and/or voltage-controlled oscillator (VCO) tanks. The first of these is generally difficult to integrate, although the loop filter and tank circuitry are easily realized in integrated form. In addition, both architectures use coarse quantization (1-2 bits) in the signal path, with a modest AGC being required in the 2-bit case. Such coarse quantization is acceptable because of the large processing gain of the GPS signal combined with its less-than-unity received SNR. In fact, there is only a 3-dB loss associated with the use of one bit, when compared to fine quantization. If two bits are used, the loss is only about 0.7dB [58].

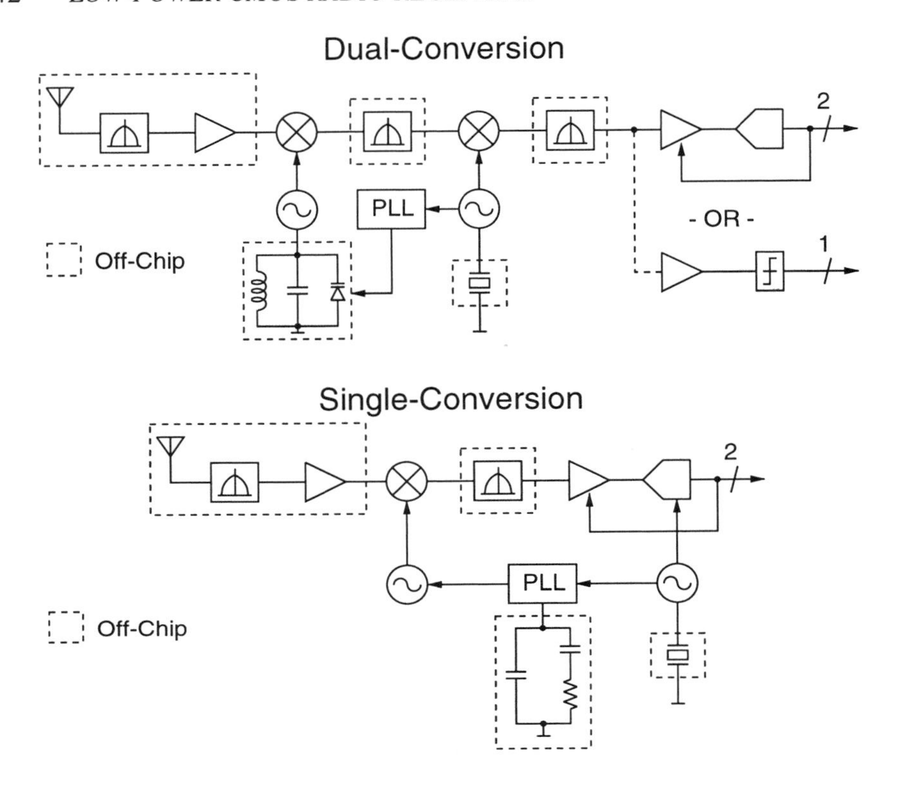

Figure 3.2. Typical GPS receiver architectures.

The clear disadvantage of these architectures is that a number of off-chip components are required. The key barrier to integration is the need for high-frequency (≈100-MHz) off-chip IF filters.

3. OPPORTUNITIES FOR A LOW-IF ARCHITECTURE

One alternative would be to implement the receiver with a low intermediate frequency architecture. This architecture suffers from the well-documented problem of limited image rejection due to the need for stringent matching of in-phase (I) and quadrature (Q) channels [55] [56]. This limitation makes the low-IF approach unsuitable for many applications. However, when we examine the GPS signal spectrum, an opportunity emerges.

In consumer applications, where the C/A code main lobe is of primary concern, we can take advantage of the narrow main lobe and relatively wide channel in a low-IF implementation. This concept is illustrated in Figure 3.3, where the L1 band has been downconverted to an intermediate frequency of 2MHz. This choice of IF causes the image signal to lie within the GPS band.

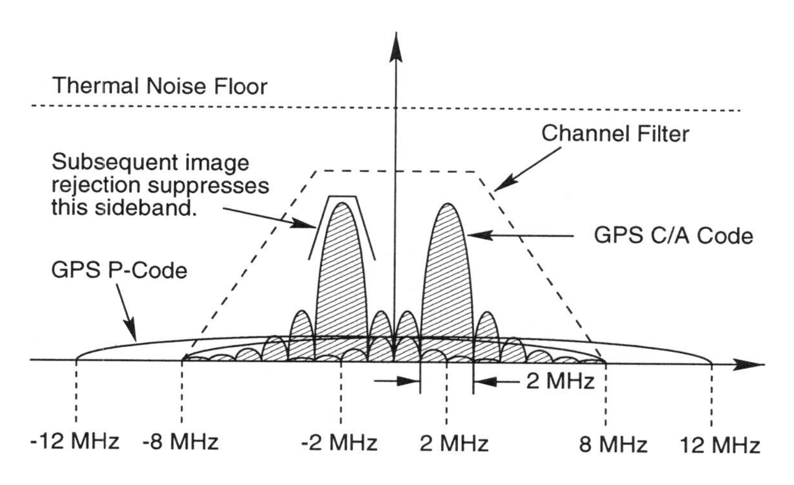

Figure 3.3. The GPS L1 band signal spectrum when downconverted to a 2-MHz intermediate frequency.

Thus, the image signal power is always comparable to the desired signal power and the required image rejection is only about 10dB, which is easily attained with ordinary levels of component matching. In addition, the spectrum from 3MHz to 8MHz can be used as a transition band for the active IF filter. This permits a reduction of the required filter order that will lead to improved dynamic range for a given filter power consumption.

These considerations make the low-IF architecture an attractive choice for a highly integrated GPS receiver.

4. GPS RECEIVER ARCHITECTURE

A detailed block diagram of a low-IF GPS receiver is shown in Figure 3.4. The complete analog signal path is implemented, including the LNA, mixers, I and Q local oscillator (LO) drivers, IF amplifiers (IFAs), active filters, limiting amplifiers (LAs) and 1-bit analog-to-digital (A/D) converters. Additionally included is an on-chip PLL comprising a VCO, loop-filter, charge pump and phase detectors. The prescaler is eliminated in favor of *aperture phase detectors* (APDs) which only operate at the reference rate, thus reducing power consumption and switching noise. The PLL is described in detail elsewhere [59].

Most of the receiver is biased with two on-chip bandgap references, with the exception of the LNA and the I and Q LO drivers. The LNA is biased with a separate self-referenced constant-g_m bias network to eliminate any possible

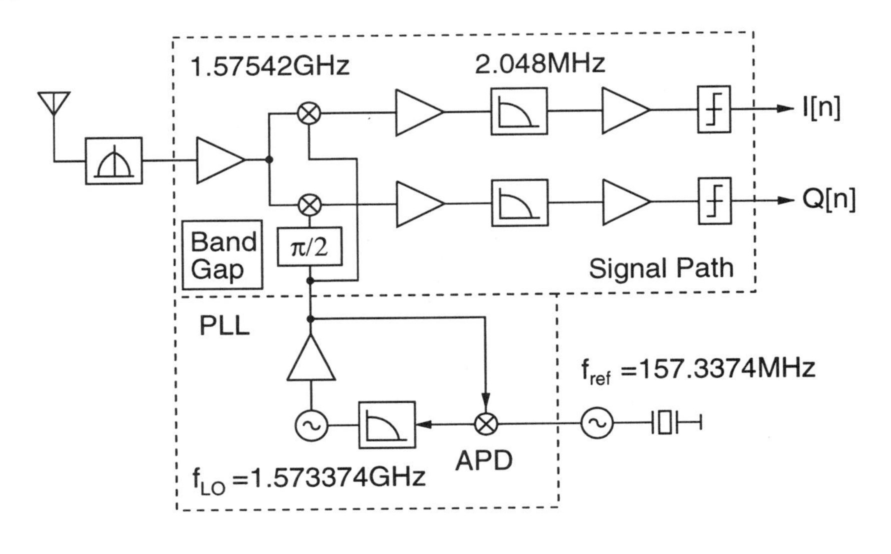

Figure 3.4. Block diagram of the CMOS GPS receiver.

interaction with other blocks through parasitic bias coupling and to stabilize its gain and input match. Similarly, the I and Q drivers are biased by another constant-g_m network for better regulation of the I and Q phase and amplitude.

4.1 IMAGE NOISE CANCELLATION

As in the familiar Weaver SSB modulator [36], the low-IF architecture depends on image cancellation to suppress the noise of the unwanted sideband. In cases where the image consists entirely of noise, the cancellation depends on the cross-correlation of the noise signals in the two channels. However, the limiters in the signal path will reduce the cross-correlation of the two noise processes, thereby drastically reducing the amount of cancellation. An analysis of this effect is presented in detail in Appendix A. This reduction in cross-correlation leads one to ask whether an image reject architecture makes sense when only 1-bit quantization is used in the I and Q signal paths.

When a low intermediate frequency is used, the noise powers in the signal band and the image band are equal. In this special case, it can be shown that the noise signals at the outputs of the limiters are uncorrelated so that subsequent downconversion and summation leads to a 3-dB SNR improvement compared to the SNR in each channel. Because this is the same improvement provided by an image-reject architecture with fine quantization, we conclude that, in this special case, the benefit of image noise rejection can be achieved despite coarse quantization in the signal path. A detailed mathematical analysis of

Table 3.1. Receiver Gain Plan

Specification	LNA	Mixer	IFA	Filter	LA
Avail. Gain (dB)	16	–4	16	–3	78
Output Z (kΩ)	0.4	0.4	2	1	60
Noise Figure (dB)	2.4	6	7	18	8
OIP3 (dBm)	7	5	10	0	–
Total NF (dB)	2.4	2.5	2.8	2.9	2.9
ONoise (dBm)	–94	–98	–81	–84	–
For P_s=–53dBm:					
OIM3 (dBm)	–125	–127	–95	–84	–
SNR (dB)	57	57	56	56	–
SDR (dB)	88	86	70	56	–

the cross-correlation between the two channels at the limiter outputs is also presented in Appendix A for completeness.

4.2 RECEIVER GAIN PLAN

For the architecture of Figure 3.4, the on-chip active filter has the narrowest dynamic range of all the signal-path elements. The active circuitry used to implement the inductive elements in the filter contributes noise and distortion, making the dynamic range much narrower than an equivalent off-chip passive filter. In particular, the noise figure of the filter is quite high (about 18dB). Because noise performance is critical in the GPS system, the receiver gain plan is tailored to suppress the active filter's contribution to the receiver noise figure.

To achieve acceptable noise performance without sacrificing too much dynamic range, the amplifier stages preceding the filter provide just enough power gain to manage the high noise figure of the filter. Table 3.1 shows the distribution of gain, noise figure and third-order intermodulation distortion throughout the receiver. From this table, it is clear that the on-chip filter is the dynamic range limiting block because it has the largest noise figure and the smallest output third-order intercept point (OIP3). Because of the dynamic range limitation imposed by the active filter, the filter design problem is treated in detail in Chapter 6.

Table 3.1 also demonstrates that the system noise figure is dominated by the noise figure of the low-noise amplifier. For the system noise figure to be less than 3dB, the LNA noise figure must be less than 2.5dB. It is challenging to achieve this LNA noise figure in a 0.5μm CMOS technology at 1.6GHz without consuming an unacceptable amount of power. Accordingly, Chapter 4 considers how to design the LNA for minimum noise figure with a given power consumption.

The frequency mixer is another important block in this architecture. It must have a relatively low noise figure and must not limit the linearity of the receiver. Note that the mixer in Table 3.1 actually has a conversion *loss* of about 4dB. By allowing for conversion loss in the architecture, a simple pair of voltage switches can be used to implement the mixer. This choice has the welcome benefit that the mixer core consumes no static power. The design of the mixer is the subject of Chapter 5.

Finally, the gain plan table gives us an indication of the peak SFDR of the receiver system. At –53-dBm available source power, the third-order intermodulation (IM3) products at the filter output have a power that is approximately equal to the noise power in a 2-MHz bandwidth. Thus, the signal-to-noise ratio is equal to the signal-to-distortion ratio. This is the condition for peak spurious-free dynamic range (SFDR), which is about 56dB.

5. SUMMARY

In summary, this chapter has presented a new architecture for a GPS receiver that is particularly well suited for integration. By taking advantage of the structure of the GPS L1-band, a low-IF receiver benefits from relaxed image rejection requirements and relaxed filter specifications. With relaxed filter order, an active filter can be used without severely compromising dynamic range. Nonetheless, the filter is still the dynamic-range limiting receiver block and must be given special attention to ensure optimized system performance.

The chapters that follow examine the LNA, mixer and filter problems in detail. Throughout, the unifying goal is the implementation of a low-power receiver system. Thus, architectural and design decisions are driven primarily by the need for *power-efficiency* in every receiver block. We begin by considering the power-constrained optimization of noise performance of the front-end low-noise amplifier.

Chapter 4

LOW-NOISE AMPLIFICATION IN CMOS AT RADIO FREQUENCIES

The first block in most wireless receivers is the low-noise amplifier (LNA). It is responsible for providing signal amplification while not degrading signal-to-noise ratio, and its noise figure sets a lower bound on the noise figure of the entire system. It is sobering to note that received GPS signal power levels at the antenna are around -130dBm and this low level degrades further in the presence of physical obstructions such as buildings and trees. Hence, a good amplifier is also critical for enabling robust performance in obstructed environments.

One possible threat to low noise operation is the well-documented, but relatively unappreciated, excess thermal noise exhibited by sub-micron CMOS devices [2][60][61][62]. This noise is believed to arise from hot electron effects in the presence of high electric fields. Despite this excess noise, recent work has demonstrated the viability of CMOS low noise amplifiers (LNA's) at frequencies around 900MHz [63][64][65].

To provide some background, Section 1. presents a review of recent LNA work in various technologies in the 900MHz–2GHz frequency range. A thorough mathematical treatment of the LNA architecture that we have chosen is presented in Section 2.. We consider in particular the effect of induced gate noise in CMOS, which is rarely cited but nonetheless of fundamental importance in establishing the limits of achievable noise performance. Section 3. discusses noise figure optimization techniques that permit selection of device geometries to maximize noise performance for a specified gain or power dissipation. In addition, numerical examples, employing the analytical techniques developed in this chapter, illustrate some of the salient features of the LNA architecture.

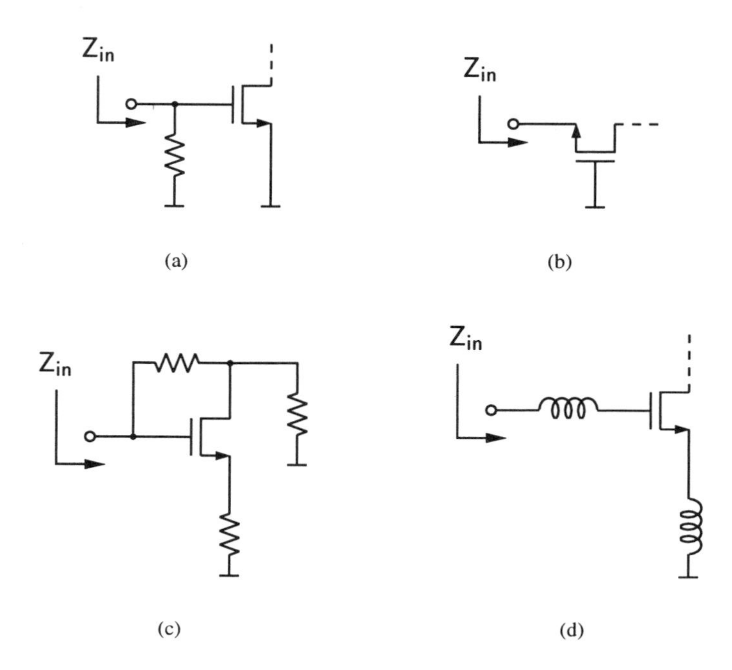

Figure 4.1. Common LNA Architectures. (a) Resistive Termination. (b) $1/g_m$ Termination. (c) Shunt-Series Feedback. (d) Inductive Degeneration.

1. RECENT LNA RESEARCH

Many authors have investigated LNA techniques in the 900MHz–2GHz frequency range. Table 4.1 summarizes the results of several recent studies dating from 1991–1998. This table has representative results from various process technologies and architectures. While the literature is full of examples of LNA work in GaAs and bipolar technologies, there are relatively few examples of CMOS studies. In addition, despite a long history of LNA work in GaAs and bipolar technologies, these papers report a wide variety of noise figures, power dissipations, and gains. The remarkable spread in published results seems to suggest that a systematic basis for the design of these amplifiers has not been elucidated. However, by examining these results from an architectural viewpoint, some order emerges.

In the design of low noise amplifiers, there are several common goals. These include minimizing the noise figure of the amplifier, providing gain with sufficient linearity — typically measured in terms of the third-order intercept point, IP3 — and providing a stable 50Ω input impedance to terminate a filter or an unknown length of transmission line which delivers signal from the

Table 4.1. Summary of Recent LNA Results

Ref.	NF (dB)	Gain (dB)	IP3/-1dB[a] (dBm)	Power (mW)	f_0 (GHz)	Arch.	Technology	Year
[66]	6.0[b]	14	na / na	7	0.75	R-Trm.	2μm CMOS	1993
[63]	2.2	15.6	12.4 / na	20	0.9	L-Deg.	0.5μm CMOS	1996
[65]	7.5	11.0	na / na	36	0.9	SSFB	1μm CMOS	1996
[64]	3.5	22	na / na	27	0.9	$1/g_m$	1μm CMOS	1996
[67]	5.3	20	11 / na	41	0.9	$1/g_m$	0.9μm CMOS	1997
[68]	2.2	17	na / na	1.8	0.9	L-Deg.	0.5μm CMOS	1997
[69]	1.9	16	9 / -4	22	0.9	L-Deg.	0.25μm CMOS	1998
[70]	3.3	9	10 / na	10	0.9	L-Deg.	0.5μm CMOS	1998
[71]	2.7	28	na / 8.5	208	1.6	SSFB	MESFET	1992
[72]	2.2	17.4	na / na	10	1.6	L-Deg.	1μm MESFET	1992
	2.2	19.6	6 / -3	10	1.0	L-Deg.	1μm MESFET	
[73]	2.0	12.2	5.1 / na	2	1.9	L-Deg.	0.3μm MESFET	1993
[74]	1.5	14.5	11.2 / -1.1	12	1.9	L-Deg.	1μm MESFET	1993
[75]	2.5	11.5	9 / na	14	1.6	L-Deg.	0.3μm MESFET	1994
[76]	5.7	7.8	23.9 / 11	115	1.0	SSFB	GaAs HBT	1991
[77]	2.2	16	6 / -4	40	0.9	L-Deg.	BiCMOS	1994
[78]	2.9	17.5	na / na	480	1.0	$1/g_m$ & SSFB	GaAs HBT	1994

[a] IP3 / -1dB compression point are output-referred.
[b] Neglects contribution of termination resistors. See text for discussion.

antenna to the amplifier. A good input match is particularly critical when a preselect filter precedes the LNA because such filters are often sensitive to the quality of their terminating impedances. The additional constraint of low power consumption which is imposed in *portable* systems further complicates the design process.

With these goals in mind, we will first focus on the requirement of providing a controlled input impedance. The architectures in Table 4.1 can be divided into four distinct approaches, illustrated in simplified form in Figure 4.1. Each of these architectures may be used in a single-ended form (as shown), or in a differential form. Note that differential forms require the use of a balun or similar element to transform the single-ended signal from the antenna into a differential signal. Practical baluns introduce extra loss which adds directly to the noise figure of the system.

The first technique uses *resistive termination* of the input port to provide a 50Ω impedance. This approach is used in its differential form by *Chang et al.* [66], for example. Unfortunately, the use of real resistors in this fashion has a deleterious effect on the amplifier's noise figure. The noise contribution of the terminating resistors is neglected in that work because an antenna would be mounted directly on the amplifier, perhaps obviating the need for input matching. Hence, the reported noise figure of 6dB corresponds to a hypothetical "terminationless" amplifier.

In general, however, the LNA is driven by a source that is located some distance away, and one must account for the influence of the terminating resistor. Specifically, we require that the amplifier possess a reasonably stable input impedance of approximately 50Ω. To evaluate the efficacy of simple resistive input termination, suppose that a given LNA employing resistive termination has an available power gain of G_a and an available noise power at the output $P_{na,i}$ due to internal noise sources only; $P_{na,i}$ is, to first order, independent of the source impedance. Then, the noise figure is found to be[1]

$$\begin{aligned} F &\triangleq \frac{Total\ output\ noise}{Total\ output\ noise\ due\ to\ the\ source} \\ &= 1 + \frac{P_{na,i} + kTBG_a}{kTBG_a} = 2 + \frac{P_{na,i}}{kTBG_a} \end{aligned} \tag{4.1}$$

where B is the bandwidth over which the noise is measured. When the amplifier termination is removed, the noise figure expression becomes, approximately,

$$F = 1 + \frac{P_{na,i}}{4kTBG_a} \tag{4.2}$$

where we have assumed a high input impedance relative to the source. From (4.1) and (4.2), we may surmise that a "terminationless" amplifier with a 6dB noise figure would likely possess an 11.5dB noise figure with the addition of the terminating resistor. Two effects are responsible for this sharp degradation in noise figure. First, the added resistor contributes its own noise to the output equal to the contribution of the source resistance. This additional noise results in a factor of two difference in the first terms of (4.1) and (4.2). Second, the input is attenuated, leading to the factor of four difference in the second terms of (4.1) and (4.2). The large noise penalty resulting from these effects therefore makes this architecture unattractive for the more common situation where a good input termination is desired.

A second architectural approach, shown in Figure 4.1(b), uses the source or emitter of a common-gate or common-base stage as the input termination.

[1] Evaluated at $T = 290K$.

A simplified analysis of the $1/g_m$-termination architecture, assuming matched conditions, yields the following lower bounds on noise figure for the cases of bipolar and CMOS amplifiers:

Bipolar:	$F = \frac{3}{2} = 1.76dB$
CMOS:	$F = 1 + \frac{\gamma}{\alpha} \geq \frac{5}{3} = 2.2dB$

where

$$\alpha \triangleq \frac{g_m}{g_{d0}}. \tag{4.3}$$

In the CMOS expressions, γ is the coefficient of channel thermal noise, g_m is the device transconductance, and g_{d0} is the zero-bias drain conductance. For long-channel devices, $\gamma = 2/3$ and $\alpha = 1$. The bipolar expression neglects the effect of base resistance in bipolar devices, while the value of 2.2dB in the CMOS expression neglects both short-channel effects ($\alpha < 1$) and excess thermal noise due to hot electrons ($\gamma > 2/3$). Indeed, for short-channel MOS devices, γ can be significantly greater than one, and α will be less than one. Accordingly, the minimum theoretically achievable noise figures tend to be around 3dB or greater in practice for this architecture.

Figure 4.1(c) illustrates yet another topology, which uses resistive shunt and series feedback to set the input and output impedances of the LNA. This approach is taken in [71], [76] and as the second stage in [78]. It is evident from Table 4.1 that amplifiers using shunt-series feedback often have significantly higher power dissipation compared to narrowband amplifiers with similar noise performance. Intuitively, the higher power is partially due to the fact that shunt-series amplifiers of this type are naturally broadband, and hence techniques which reduce the power consumption through L-C tuning are not applicable. For GPS applications, a broadband front end is not required and it is additionally desirable to make use of narrowband techniques to reduce power and reject out-of-band interfering signals and noise. Furthermore, the shunt-series architecture requires on-chip resistors of reasonable quality, which are generally not available in CMOS technologies. For these reasons, the shunt-series feedback approach is not pursued in this work.

The fourth architecture, and the one that we have used in this design, employs inductive source or emitter degeneration as shown in Figure 4.1(d) to generate a real term in the input impedance [79]. Tuning of the amplifier input becomes necessary, making this a narrow band approach. However, this requirement is not a limitation for a GPS receiver.

Note that inductive source degeneration is the most prevalent method used for GaAs MESFET amplifiers. It has also been used in CMOS amplifiers recently at 900MHz [63]. As we will see, the proliferation of this architecture

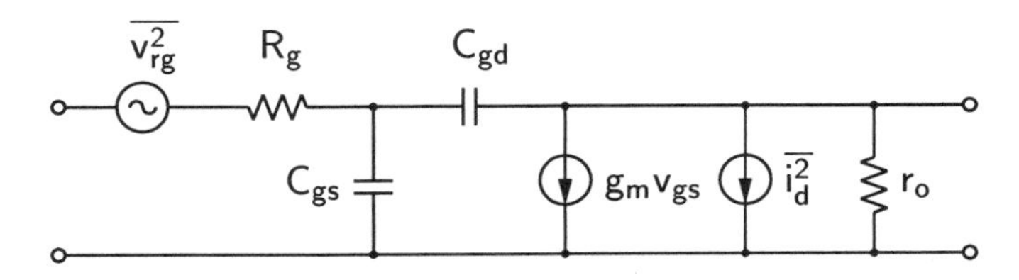

Figure 4.2. The standard CMOS noise model.

is no accident; it offers the possibility of achieving the best noise performance of *any* architecture.

2. LNA ARCHITECTURAL ANALYSIS

We will now pursue a careful analysis of the architecture in Figure 4.1(d) to establish clearly the principle of operation and the limits on noise performance. A brief review of the standard CMOS noise model will facilitate the analysis.

2.1 STANDARD MOS NOISE MODEL

The standard CMOS noise model is shown in Figure 4.2. The dominant noise source in CMOS devices is channel thermal noise. This source of noise is commonly modeled as a shunt current source in the output circuit of the device. The channel noise is white with a power spectral density given by

$$\frac{\overline{i_d^2}}{\Delta f} = 4kT\gamma g_{d0} \tag{4.4}$$

where g_{d0} is the zero-bias drain conductance of the device and γ is a bias-dependent factor that, for long-channel devices, satisfies the inequality

$$\frac{2}{3} \leq \gamma \leq 1. \tag{4.5}$$

The value of $2/3$ holds when the device is saturated, and the value of 1 is valid when the drain-source voltage is zero. For short-channel devices, however, γ does not satisfy Equation (4.5). In fact, γ *can be much greater than 2/3 for short-channel devices operating in saturation* [2][62]. For 0.7-μm channel lengths, γ may be as high as 2–3, depending on bias conditions [2].

This excess noise may be attributed to the presence of hot electrons in the channel. The high electric fields in sub-micron MOS devices cause the electron temperature, T_e, to exceed the lattice temperature. The excess noise due to carrier heating was anticipated by van der Ziel as early as 1970 [80].

An additional source of noise in MOS devices is the noise generated by the distributed gate resistance [81]. This noise source can be modeled by a series resistance in the gate circuit and an accompanying white noise generator. By interdigitating the device, the contribution of this source of noise can be reduced to insignificant levels. For noise purposes, the effective gate resistance is given by [82]

$$R_g = \frac{R_\square W}{3n^2 L} \tag{4.6}$$

where $R_\square$ is the sheet resistance of the polysilicon, W is the total gate width of the device, L is the gate length, and n is the number of gate fingers used to lay out the device. The factor of $1/3$ arises from a distributed analysis of the gate, assuming that each gate finger is contacted only at one end. By contacting at *both* ends, this term reduces to $1/12$. In addition, this expression neglects the interconnect resistance used to connect the multiple gate fingers together. The interconnect can be routed in a metal layer that possesses significantly lower sheet resistance, and hence is easily rendered insignificant.

Though playing a role similar to that of base resistance in bipolar devices, the gate resistance is much less significant in CMOS because it can be minimized through interdigitation without the need for increased power consumption, unlike its bipolar counterpart. Its significance is further reduced in silicided CMOS processes which possess a greatly reduced sheet resistance, $R_\square$.

An additional source of noise in CMOS devices is the back-gate epitaxial resistance [67], which can result in an apparent increase in γ, the coefficient of drain noise. To evaluate the magnitude of the epitaxial resistance noise, we can model the epitaxial layer as a resistance in series with the bulk terminal of the device [83]. There is a noise voltage associated with this resistance which, together with the drain current noise of the device, produces a total drain current noise of

$$\frac{\overline{i_d^2}}{\Delta f} = 4kT\left\{\gamma g_{d0} + g_{mb}^2 R_{epi}\right\} = 4kT\gamma_{eff}\, g_{d0} \tag{4.7}$$

where

$$\gamma_{eff} \approx \gamma + \frac{g_{mb}^2 R_{epi}}{g_{d0}}. \tag{4.8}$$

For the 0.5-μm technology used in this work, and for the particular device size and layout geometry used in the LNA presented in Chapter 7,

$$0.09 \leq \frac{g_{mb}^2 R_{epi}}{g_{d0}} \leq 0.2. \tag{4.9}$$

The lower bound uses the approach outlined in [83] for estimating R_{epi}, while the upper bound uses a much more conservative approach based on a trapezoidal approximation for the epitaxial spreading resistance [84], assuming that

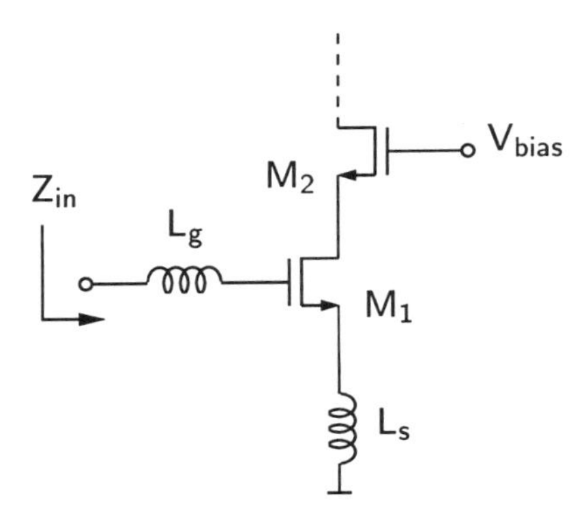

Figure 4.3. Common-source input stage.

substrate contacts are distant from the device. With closely-spaced substrate contacts, this number will be reduced even further. Thus, the epitaxial resistance is of secondary importance in this case and will be ignored in the analysis that follows.

2.2 LNA ARCHITECTURE

We now proceed to the analysis of the LNA architecture using the standard CMOS noise model. Figure 4.3 illustrates the input stage of the LNA. A simple analysis of the input impedance shows that

$$\begin{aligned} Z_{in} &= s(L_s + L_g) + \frac{1}{sC_{gs}} + \left(\frac{g_{m1}}{C_{gs}}\right) L_s \\ &\approx \omega_T L_s \qquad (at\ resonance). \end{aligned} \tag{4.10}$$

If the effect of the gate-to-drain overlap capacitance is included, then, at resonance

$$Z_{in} \approx \frac{\omega_T L_s}{1 + 2\frac{C_{gd}}{C_{gs}}} = \omega_{T,eff} L_s. \tag{4.11}$$

At the series resonance of the input circuit, the impedance is purely real and proportional to L_s. By choosing L_s appropriately, this real term can be made equal to 50Ω. For example, if f_T is 10GHz, a 50Ω impedance requires only 800pH for L_s. This small amount of inductance can easily be obtained with a single bondwire or on-chip spiral inductor. The gate inductance L_g is used to set the resonance frequency once L_s is chosen to satisfy the criterion of a 50Ω input impedance.

The noise figure of the LNA can be computed by analyzing the circuit shown in Figure 4.4. In this circuit, R_l represents the series resistance of the inductor

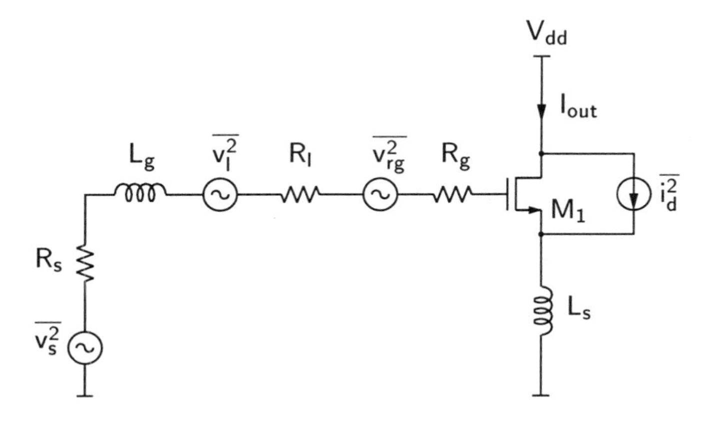

Figure 4.4. Equivalent circuit for input stage noise calculations.

L_g, R_g is the gate resistance of the NMOS device, and $\overline{i_d^2}$ represents the channel thermal noise of the device. Analysis based on this circuit neglects the contribution of subsequent stages to the amplifier noise figure. This simplification is justifiable provided that the first stage possesses sufficient gain, and permits us to examine in detail the salient features of this architecture. Note that the overlap capacitance, C_{gd}, has also been neglected in the interest of simplicity. The use of a cascoded first stage helps to ensure that this approximation will not introduce serious errors.

Recall that the noise figure for an amplifier is defined as[2]

$$F \triangleq \frac{Total\ output\ noise}{Total\ output\ noise\ due\ to\ the\ source}. \tag{4.12}$$

To evaluate the output noise when the amplifier is driven by a 50Ω source, we first evaluate the transconductance of the input stage. With the output current proportional to the voltage on C_{gs}, and noting that the input circuit takes the form of a series-resonant network,

$$\begin{aligned} G_m &= g_{m1} Q_{in} = \frac{g_{m1}}{\omega_0 C_{gs}(R_s + \omega_T L_s)} \\ &= \frac{\omega_T}{\omega_0 R_s \left(1 + \frac{\omega_T L_s}{R_s}\right)} = \frac{\omega_T}{2\omega_0 R_s} \end{aligned} \tag{4.13}$$

where Q_{in} is the effective Q of the amplifier input circuit. In this expression, which is valid at the series resonance ω_0, R_l and R_g have been neglected relative

[2]Evaluated at a source temperature $T = 290K$.

to the source resistance, R_s. Perhaps surprisingly, the transconductance of this circuit at resonance is *independent* of g_{m1} (the device transconductance) as long as the resonant frequency is maintained constant. If, at constant bias voltages, the width of the device is adjusted, the transconductance of the *stage* will remain the same as long as L_g is adjusted to maintain a fixed resonant frequency. This result is intuitively satisfying, for as the gate width (and thus g_{m1}) is reduced, C_{gs} is also reduced, resulting in an increased Q_{in} such that the product of g_{m1} and Q_{in} remains fixed.

Using (4.12), the output noise power density due to the 50Ω source is

$$S_{a,src}(\omega_0) = S_{src}(\omega_0)G_{m,eff}^2 = \frac{4kT\omega_T^2}{\omega_0^2 R_s \left(1 + \frac{\omega_T L_s}{R_s}\right)^2}. \tag{4.14}$$

In a similar fashion, the output noise power density due to R_l and R_g can be expressed as

$$S_{a,R_l,Rg}(\omega_0) = \frac{4kT(R_l + R_g)\omega_T^2}{\omega_0^2 R_s^2 \left(1 + \frac{\omega_T L_s}{R_s}\right)^2}. \tag{4.15}$$

Equations (4.14) and (4.15) are also valid only at the series resonance of the circuit.

The dominant noise contributor internal to the LNA is the channel current noise of the first MOS device. Recalling the expression for the power spectral density of this source from (4.4), one can derive that the output noise power density arising from this source is

$$S_{a,i_d}(\omega_0) = \frac{\frac{\overline{i_d^2}}{\Delta f}}{\left(1 + \frac{\omega_T L_s}{R_s}\right)^2} = \frac{4kT\gamma g_{d0}}{\left(1 + \frac{\omega_T L_s}{R_s}\right)^2}. \tag{4.16}$$

The total output noise power density is the sum of (4.14)–(4.16). Assuming a 1Hz bandwidth and substituting these into Equation (4.12) yields

$$F = 1 + \frac{R_l}{R_s} + \frac{R_g}{R_s} + \gamma g_{d0} R_s \left(\frac{\omega_0}{\omega_T}\right)^2 \tag{4.17}$$

which is the noise figure of the LNA.

This equation for noise figure reveals several important features of this LNA architecture. Note that the dominant term in (4.17) is the last term, which arises from channel thermal noise. Surprisingly, this term is *proportional* to g_{d0}. So, according to this expression, by reducing g_{d0} without modifying ω_T, we can simultaneously improve noise figure *and* reduce power dissipation. We can achieve this result by scaling the width of the device while maintaining constant bias voltages on its terminals and leaving the channel length unchanged. This scaling is consistent with the condition of constant ω_T, which depends only on the bias *voltages* on the device.

Recall, however, that this expression assumes that the amplifier is operated at the series resonance of its input circuit. So, a reduction in g_{d0} (and, hence in C_{gs}) must be compensated by an increase in L_g to maintain a constant resonant frequency. So, better noise performance and reduced power dissipation can be obtained by increasing the Q of the input circuit resonance.

By applying device scaling in this fashion to improve noise performance, the linearity of the amplifier will tend to degrade due to increased signal levels across C_{gs}. However, short-channel MOS devices operating in velocity saturation have a relatively constant transconductance with sufficient gate overdrive voltage. This property is one advantage of implementing LNA's with MOS devices.

A second important feature in (4.17) is the inverse dependence on ω_T^2. Continued improvements in technology will therefore naturally lead to improved noise performance at a given frequency of operation.

Careful examination of (4.17) reveals a curious feature, however. Although finite inductor Q's will limit the amount of improvement practically available through device scaling, Equation (4.17) does not predict a *fundamental* minimum for F. The implication is that a 0dB noise figure may be achieved with zero power dissipation, and this result simply cannot be true. Yet, the expression follows directly from the MOS noise model that we have assumed.

The conclusion can only be that our noise model is incomplete.

2.3 EXTENDED MOS NOISE MODEL

To understand the fundamental limits on noise performance of this architecture, we must turn our attention to induced gate current noise in MOS devices. Although absent from most texts on CMOS circuit design, gate noise is given detailed treatment by van der Ziel and others [80] [1] [85] [86] [87].

Figure 4.5 shows the cross-section of a MOS device. If the device is biased so that the channel is inverted, fluctuations in the channel charge will induce a physical current in the gate due to capacitive coupling. This noise current can be (and has been) measured [85], but it is not included in the simple MOS noise model that we have used in the previous section.

A companion effect that occurs at very high frequencies arises due to the "distributed" nature of the MOS device. At frequencies approaching ω_T, the gate impedance of the device exhibits a significant phase shift from its purely capacitive value at lower frequencies. This shift can be accounted for by including a real conductance, g_g, in the gate circuit. Note that this conductance

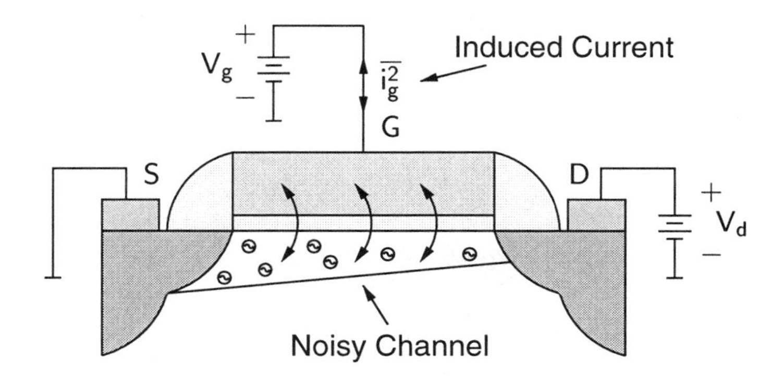

Figure 4.5. Induced gate effects in MOS devices.

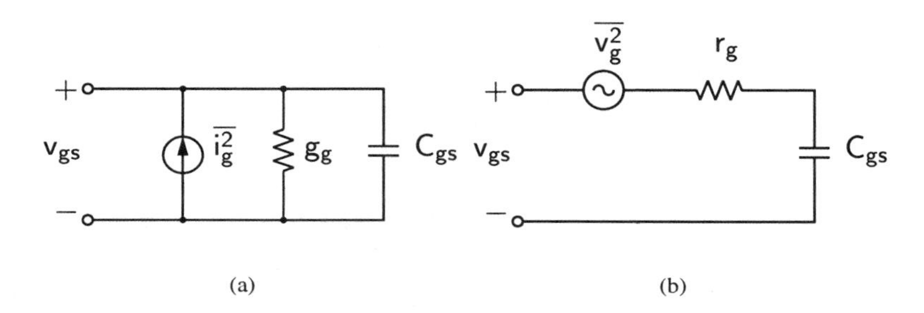

Figure 4.6. Revised gate circuit model including induced effects. (a) Standard representation, as found in [1]. (b) The equivalent, but more intuitive, Thévenin representation.

is distinct from the polysilicon resistance and is also distinct from the real term that occurs due to interaction of C_{gd} with g_m[3].

A simple gate circuit model that includes both of these effects is shown in Figure 4.6(a). A shunt noise current $\overline{i_g^2}$ and a shunt conductance g_g have been added. Mathematical expressions for these sources are [1][4]

$$\frac{\overline{i_g^2}}{\Delta f} = 4kT\delta g_g \tag{4.18}$$

$$g_g = \frac{\omega^2 C_{gs}^2}{5 g_{d0}} \tag{4.19}$$

[3]A real conductance with a form similar to g_g is generated in cascoded amplifiers due to the feedback provided by C_{gd}. This distinctly different effect is also significant at frequencies approaching ω_T.

[4]Our notation differs slightly from that found in [1], in which β is used in place of δ. The use of δ avoids confusion in cases where β represents $\mu_n C_{ox} W/L$, as is the practice in some texts on MOS devices.

where δ is the coefficient of gate noise, classically equal to $4/3$ for long-channel devices. Equations (4.18) and (4.19) are valid when the device is operated in saturation.

Some observations about (4.18) and (4.19) are warranted. Note that the expression for the gate noise power spectral density takes a form similar to that of Equation (4.4), which describes the drain noise power spectral density. However, in the gate noise expression, g_g is proportional to ω^2, and hence the gate noise is *not* a white noise source. Indeed, it is better described as a "blue" noise source due to its monotonically increasing power spectral density. It seems mysterious that the gate and drain noise terms have different types of power spectra, given their common progenitor. The mystery is somewhat artificial, however, because the circuit of Figure 4.6(a) can be cast into an equivalent Thévenin representation as shown in Figure 4.6(b) where

$$\frac{\overline{v_g^2}}{\Delta f} = 4kT\delta r_g \tag{4.20}$$

$$r_g = \frac{1}{5g_{d0}}. \tag{4.21}$$

We observe that v_g is now a *white* noise source proportional to a constant resistive term, r_g. This formulation of the gate circuit seems more intuitively appealing because the frequency dependence has been removed for *both* terms. Figures 4.6(a) and 4.6(b) are interchangeable for frequencies where the Q of C_{gs} is sufficiently large, i.e.,

$$Q_{C_{gs}} = \frac{5g_{d0}}{\omega C_{gs}} \gg 1 \tag{4.22}$$

or, equivalently,

$$\omega \ll \frac{5g_{d0}}{C_{gs}} = \frac{5\omega_T}{\alpha} \tag{4.23}$$

where α was defined in Equation (4.3) and is always less than one. This condition is automatically satisfied in all cases of practical interest.

In addition, we can expect the coefficient of gate noise, δ, to exhibit a dependence on electric field just as its counterpart, γ. To our knowledge, though, there are no published studies of the high-field behavior of δ.

The presence of gate noise complicates the analysis of F significantly. The gate noise is *partially correlated* with the drain noise, with a complex correlation coefficient given by [1]

$$c = \frac{\overline{i_g i_d^*}}{\sqrt{\overline{i_g^2}\,\overline{i_d^2}}} \approx 0.395j \tag{4.24}$$

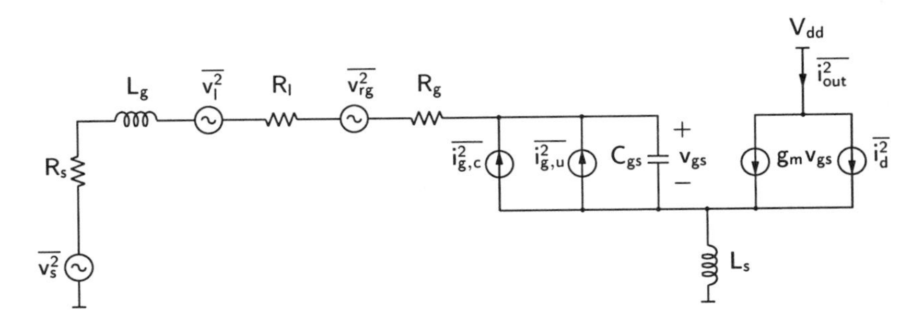

Figure 4.7. Revised small-signal model for LNA noise calculations.

where the value of $0.395j$ is exact for long-channel devices. The correlation can be treated by expressing the gate noise as the sum of two components, the first of which is fully correlated with the drain noise, and the second of which is uncorrelated with the drain noise. Hence, the gate noise is re-expressed as

$$\frac{\overline{i_g^2}}{\Delta f} = \underbrace{4kT\delta g_g(1-|c|^2)}_{Uncorrelated} + \underbrace{4kT\delta g_g|c|^2}_{Correlated}. \tag{4.25}$$

Because of the correlation, special attention must be paid to the reference polarity of the correlated component. The value of c is positive for the polarity shown in Figure 4.6(a).

Having established this additional source of noise in MOS devices, we are now in a position to re-evaluate the noise figure of the LNA. As we will see, the presence of gate noise establishes a lower bound on the achievable noise performance of the amplifier.

2.4 EXTENDED LNA NOISE ANALYSIS

To evaluate the noise performance of the LNA in the presence of gate noise effects, we will employ the circuit of Figure 4.7. In this circuit, we have neglected the effect of g_g under the assumption that the gate impedance is largely capacitive at the frequency of interest. Equation (4.23) specifies the condition under which this approximation holds. The gate noise has been subdivided into two parts. The first, $\overline{i_{g,c}^2}$, represents the portion of the total gate noise that is correlated with the drain noise. The second, $\overline{i_{g,u}^2}$, represents the portion that is uncorrelated with the drain noise.

With the revised small-signal model in mind, we can derive the noise figure of the LNA. A close examination of Figure 4.7 allows us to anticipate the result of our analysis. As the Q of the input circuit is increased from zero, the noise figure will tend to improve in accordance with the earlier expression for F. However, the impedance at the gate of the device increases simultaneously, and hence the gate current noise will begin to dominate at some point. A minimum noise figure will thus be achieved for a particular input Q.

To analyze the circuit mathematically, we can draw on Equations (4.14)–(4.16) from the previous section for the drain noise and resistive losses. However, the *amplitudes* of the correlated portion of the gate noise and the drain noise must be summed together before the *powers* of the various contributors are summed. Doing so yields a term representing the combined effect of the drain noise and the correlated portion of the gate noise

$$S_{a,i_d,i_{g,c}}(\omega_0) = \kappa S_{a,i_d}(\omega_0) = \frac{4kT\gamma\kappa g_{d0}}{\left(1+\frac{\omega_T L_s}{R_s}\right)^2} \tag{4.26}$$

where,

$$\kappa = \frac{\delta\alpha^2}{5\gamma}|c|^2 + \left[1+|c|\sqrt{\frac{\delta\alpha^2}{5\gamma}}\right]^2. \tag{4.27}$$

Note that if $\delta \to 0$, then $\kappa \to 1$ and Equation (4.26) then reduces to Equation (4.16).

The last noise term is the contribution of the uncorrelated portion of the gate noise. This contributor has the following power spectral density:

$$S_{a,i_{g,u}}(\omega_0) = \xi S_{a,i_d}(\omega_0) = \frac{4kT\gamma\xi g_{d0}}{\left(1+\frac{\omega_T L_s}{R_s}\right)^2} \tag{4.28}$$

where,

$$\xi = \frac{\delta\alpha^2}{5\gamma}(1-|c|^2)(1+Q_s^2) \tag{4.29}$$

$$Q_s = \frac{\omega_0(L_s+L_g)}{R_s} = \frac{1}{\omega_0 R_s C_{gs}}. \tag{4.30}$$

We observe that all of the noise terms contributed by the first device, M_1, are proportional to $S_{a,i_d}(\omega_0)$, the contribution of the drain noise. Hence, it is convenient to define the contribution of M_1 as a whole as

$$S_{a,M_1}(\omega_0) = \chi S_{a,i_d}(\omega_0) = \frac{4kT\gamma\chi g_{d0}}{\left(1+\frac{\omega_T L_s}{R_s}\right)^2} \tag{4.31}$$

where, after some slight simplification,

$$\chi = \kappa + \xi = 1 + 2|c|\sqrt{\frac{\delta\alpha^2}{5\gamma}} + \frac{\delta\alpha^2}{5\gamma}(1+Q_s^2). \tag{4.32}$$

With Equations (4.31) and (4.32), it is clear that the effect of induced gate noise is to modify the noise contribution of the device in proportion to χ. It follows directly that

$$F = 1 + \frac{R_l}{R_s} + \frac{R_g}{R_s} + \gamma \chi g_{d0} R_s \left(\frac{\omega_0}{\omega_T} \right)^2 \tag{4.33}$$

where χ is defined as in (4.32). By factoring out Q_s from the expression for χ, and noting that

$$g_{d0} Q_s = \frac{g_m}{\alpha} \frac{1}{\omega_0 R_s C_{gs}} = \frac{\omega_T}{\alpha \omega_0 R_s} \tag{4.34}$$

we can re-express F as

$$F = 1 + \frac{R_l}{R_s} + \frac{R_g}{R_s} + \frac{\gamma}{\alpha} \frac{\chi}{Q_s} \left(\frac{\omega_0}{\omega_T} \right). \tag{4.35}$$

To understand the implications of this new expression for F, we observe that χ includes terms which are constant and terms which are proportional to Q_s^2. It follows that (4.35) will contain terms which are *proportional* to Q_s as well as *inversely proportional* to Q_s. Therefore, a minimum F exists for a particular Q_s, as argued earlier. Selection of the optimum Q_s forms the subject of the next section.

3. LNA DESIGN CONSIDERATIONS

The analysis of the previous section can now be drawn upon in designing the LNA. Of primary interest is insight into picking the appropriate device width and bias point to optimize noise performance given specific objectives for gain and power dissipation.

To select the width of M_1, we turn to Equations (4.32) and (4.35). Note that all of the terms are well-defined in these expressions, except for γ and δ. Because γ and δ both depend on drain bias in an unspecified fashion, it is difficult to account properly for their contributions. To surmount this difficulty, we adopt the assumption that although each may be a function of bias, the *ratio* can be expected to show less variation because γ and δ will likely have similar dependence on bias, given their common progenitor. The reader is cautioned, however, that this assumption is somewhat arbitrary; it is necessary because the detailed high-field behavior of γ and δ is presently unknown. Modifications may be required once further research yields information about these coefficients. The analysis which follows is sufficiently general, however, that it can be easily adapted to accommodate new information on the high-field natures of γ and δ.

In preparation for optimizing the noise performance of the LNA, it will be useful to formulate the quantities α, ω_T, and Q_s in terms of the gate overdrive voltage of M_1.

3.1 A SECOND-ORDER MOSFET MODEL

To quantify these terms, a simple second-order model of the MOSFET transconductance can be employed which accounts for high-field effects in short-channel devices. Assume that I_d has the form [3]

$$I_d = WC_{ox}\nu_{sat}\frac{V_{od}^2}{V_{od} + L\varepsilon_{sat}} \tag{4.36}$$

with

$$V_{od} = V_{gs} - V_T \tag{4.37}$$

where C_{ox} is the gate oxide capacitance per unit area, ν_{sat} is the saturation velocity, and ε_{sat} is the velocity saturation field strength, defined as the lateral electric field for which the mobility drops to one half of its low-field value. We can differentiate this expression to determine the transconductance, yielding

$$g_m = \frac{\partial I_d}{\partial V_{gs}} = \mu_{eff}C_{ox}\frac{W}{L}V_{od}\underbrace{\left[\frac{1+\rho/2}{(1+\rho)^2}\right]}_{\alpha} \tag{4.38}$$

with the definition that

$$\rho = \frac{V_{od}}{L\varepsilon_{sat}} \tag{4.39}$$

where μ_{eff} is the field-limited electron mobility. The term in square braces is α itself.

Having established an expression for I_d, we can formulate the power consumption of the amplifier as follows,

$$P_D = V_{dd}I_d = V_{dd}WC_{ox}\nu_{sat}\frac{V_{od}^2}{V_{od} + L\varepsilon_{sat}}. \tag{4.40}$$

Note that the power dissipation is proportional to the device width, W. Another quantity which depends directly on W is Q_s, which has been specified in Equation (4.30). Combining this equation with (4.40), and noting that $C_{gs} = \frac{2}{3}WLC_{ox}$, we can relate Q_s to P_D with

$$Q_s = \frac{P_0}{P_D}\frac{\rho^2}{1+\rho} \tag{4.41}$$

where

$$P_0 = \frac{3}{2}\frac{V_{dd}\nu_{sat}\varepsilon_{sat}}{\omega_0 R_s}. \tag{4.42}$$

Note that for the purposes of our analysis, P_0 is a constant determined solely by physical technological parameters (ν_{sat} and ε_{sat}) and design target specifications (V_{dd}, ω_0, and R_s).

Another factor required in the design process is ω_T. This can also be evaluated with the help of (4.38) to be

$$\omega_T \approx \frac{g_m}{C_{gs}} = \frac{g_m}{\frac{2}{3}WLC_{ox}} = \frac{3}{2}\frac{\alpha\mu_{eff}V_{od}}{L^2} = \frac{3\alpha\rho\nu_{sat}}{L}. \tag{4.43}$$

This expression is approximate because we are neglecting C_{gd}, the gate-drain overlap capacitance. This approximation is reasonable if we assume that the LNA input stage is cascoded. Note that proportionality to α limits the ω_T that can be achieved with a given device.

3.2 NOISE FIGURE OPTIMIZATION TECHNIQUES

With the relevant quantities now defined, we can proceed to optimize the noise performance of the amplifier. In low noise amplifier design, determination of the minimum noise figure is a common and well-understood procedure. Typically, a small-signal model of the amplifier is assumed *a priori*, an expression for F is formed, and differentiation leads to a unique source impedance that optimizes noise performance. The reader is referred to [43] for an excellent treatment of the general approach. Note that the assumption of a fixed small-signal model reflects a discrete amplifier mindset in which device geometry is not under the designer's control. There is a significant distinction, however, between that type of optimization and the one which we seek to perform here. In an integrated circuit environment, the device geometry is flexible and can be incorporated into the optimization procedure. Thus, we begin by specifying a desired design parameter, such as gain or power consumption, under the condition of perfect input matching. Then, we determine the appropriate small-signal model *a posteriori* through the optimization procedure. By selecting Q_s and L_s independently, we can determine the device geometry that yields optimized noise performance with an excellent input match.

There are two approaches to this optimization problem which deserve special attention. The first assumes a fixed transconductance, G_m, for the amplifier. The second assumes a fixed power consumption. To illustrate the second approach, the expression for F in Equation (4.35) can be re-cast to make its dependence on power dissipation (P_D) explicit. It is, however, non-trivial to make the dependence on G_m explicit. Fortunately, the condition for constant G_m is equivalent to the condition of constant ω_T, as is clear from Equation (4.12). To maintain a fixed ω_T, we need only fix the value of ρ. Hence, we will reformulate F in terms of P_D and ρ to facilitate both optimizations.

We can draw on Equations (4.38), (4.41), and (4.43) and substitute into (4.35) expressions for α, Q_s, and ω_T in terms of the relative gate overdrive voltage, ρ. The result is that

$$F = 1 + \frac{\gamma\omega_0 L}{3\nu_{sat}} P(\rho, P_D) \tag{4.44}$$

in which we have neglected the contributions of the gate resistance and inductor losses to the noise figure. In this new expression, $P(\rho, P_D)$ is a ratio of two 6th-order polynomials of ρ given by

$$P(\rho, P_D) = \frac{\frac{P_D}{P_0} P_1(\rho) + \frac{P_0}{P_D} P_2(\rho)}{\rho^3 \left(1 + \frac{\rho}{2}\right)^2 (1 + \rho)} \tag{4.45}$$

with

$$P_1(\rho) = (1+\rho)^6 + 2|c|\sqrt{\frac{\delta}{5\gamma}}\left(1 + \frac{\rho}{2}\right)(1+\rho)^4 + \frac{\delta}{5\gamma}(1+\rho)^2\left(1+\frac{\rho}{2}\right)^2 \tag{4.46}$$

$$P_2(\rho) = \frac{\delta}{5\gamma}\left(1 + \frac{\rho}{2}\right)^2 \rho^4. \tag{4.47}$$

The form of Equation (4.44) suggests that optimization of F proceeds by minimizing $P(\rho, P_D)$ with respect to one of its arguments, keeping the other one fixed. The complexity of this polynomial will force us to make some simplifying assumptions when optimizing for a fixed power dissipation. Fortunately, the optimization for a fixed G_m can proceed directly from (4.45) without further simplifications.

3.2.1 FIXED G_M OPTIMIZATION

To fix the value of the transconductance, G_m, we need only assign a constant value to ρ. The appropriate value for ρ is easily determined by substituting (4.43) into the expression for G_m as found in (4.12). The result, which relates G_m to ρ, is

$$G_m = \frac{3\nu_{sat}}{2\omega_0 R_s L} \frac{\rho(1 + \frac{\rho}{2})}{(1+\rho)^2}. \tag{4.48}$$

Once ρ is determined, we can minimize the noise figure by taking

$$\frac{\partial P(\rho, P_D)}{\partial P_D} = 0 \tag{4.49}$$

which, after some algebraic manipulations, results in

$$P_{D,opt,G_m} = P_0\sqrt{\frac{P_3(\rho)}{P_1(\rho)}} = P_0 \frac{\rho^2}{1+\rho}\left[1 + 2|c|\sqrt{\frac{5\gamma}{\delta\alpha^2}} + \frac{5\gamma}{\delta\alpha^2}\right]^{-1/2}. \tag{4.50}$$

This expression gives the power dissipation which yields the best noise performance for a given G_m under the assumption of a matched input impedance. By comparing (4.50) to (4.41), we see immediately that this optimum occurs when

$$Q_s = Q_{s,opt,G_m} = \sqrt{1 + 2|c|\sqrt{\frac{5\gamma}{\delta\alpha^2}} + \frac{5\gamma}{\delta\alpha^2}} \geq 2.183. \tag{4.51}$$

Hence, the best noise performance for a given transconductance is achieved at some specific input Q. Note that the value 2.183 is valid only for long-channel devices. For short-channel lengths, where $\alpha < 1$, we can expect the optimum Q_s to be somewhat larger. Note that if we substitute Q_{s,opt,G_m} into (4.32), the sum of the second two terms (which are attributed to the presence of gate noise) exceeds unity, thus indicating that the gate current contributes *more* noise than the drain current.

By substituting (4.51) into (4.35), we determine that the minimum noise figure (neglecting inductor and gate losses) is

$$\begin{aligned} F_{min,G_m} &= 1 + \sqrt{\frac{4}{5}\delta\gamma}\left(\frac{\omega_0}{\omega_T}\right)\left\{1 + 2|c|\sqrt{\frac{\delta\alpha^2}{5\gamma}} + \frac{\delta\alpha^2}{5\gamma}\right\}^{1/2} \\ &= 1 + 1.235\sqrt{\delta\gamma}\left(\frac{\omega_0}{\omega_T}\right) \geq 1 + 1.164\left(\frac{\omega_0}{\omega_T}\right). \end{aligned} \tag{4.52}$$

Note that the coefficient 1.164 is valid only in the long-channel limit and is likely to be somewhat larger in short-channel devices due to hot-electron effects.

3.2.2 FIXED P_D OPTIMIZATION

An alternate method of optimization fixes the power dissipation and adjusts ρ to find the minimum noise figure. The expression for $P(\rho, P_D)$ is too complex in ρ to yield a closed form solution for the optimum point. However, we can adopt a simplifying assumption and check its validity by graphical comparison. If we assume that $\rho \ll 1$, then $P(\rho, P_D)$ can be simplified to

$$P(\rho, P_D) \approx \frac{\frac{P_D}{P_0}\left(1 + 2|c|\sqrt{\frac{\delta}{5\gamma}} + \frac{\delta}{5\gamma}\right) + \frac{P_0}{P_D}\frac{\delta}{5\gamma}\rho^4}{\rho^3}. \tag{4.53}$$

This expression is minimized for a fixed P_D when

$$\frac{\partial P(\rho, P_D)}{\partial \rho} = 0. \tag{4.54}$$

The solution of this equation, under the assumption that $\rho \ll 1$ is

$$\rho^2_{opt,P_D} = \frac{P_D}{P_0}\sqrt{3}\left\{1 + 2|c|\sqrt{\frac{5\gamma}{\delta\alpha^2}} + \frac{5\gamma}{\delta\alpha^2}\right\}^{1/2}. \quad (4.55)$$

By comparing (4.55) to (4.41), it is clear that this value for ρ is equivalent to an optimum Q_s of

$$\begin{aligned} Q_{s,opt,P_D} &= \sqrt{3}\left\{1 + 2|c|\sqrt{\frac{5\gamma}{\delta\alpha^2}} + \frac{5\gamma}{\delta\alpha^2}\right\}^{1/2} = \sqrt{3}Q_{s,opt,G_m} \\ &\approx 3.781. \end{aligned} \quad (4.56)$$

So, it is clear that the optimum Q_s for a fixed power dissipation is *larger* than the optimum Q_s for a fixed G_m. We can now evaluate Equation (4.32) and use the result in (4.35) to show that

$$\begin{aligned} F_{min,P_D} &= 1 + \sqrt{\frac{16}{15}\delta\gamma}\left(\frac{\omega_0}{\omega_T}\right)\left\{1 + 2|c|\sqrt{\frac{\delta\alpha^2}{5\gamma}} + \frac{\delta\alpha^2}{5\gamma}\right\}^{1/2} \\ &= 1 + 1.426\sqrt{\delta\gamma}\left(\frac{\omega_0}{\omega_T}\right) \geq 1 + 1.344\left(\frac{\omega_0}{\omega_T}\right) \end{aligned} \quad (4.57)$$

where the value of 1.344 is valid only in the long-channel limit; the value will be somewhat larger for short-channel devices in velocity saturation.

To examine the validity of our simplifying assumption that $\rho \ll 1$, the noise figure is plotted in Figure 4.8 for the two cases defined in (4.45) and (4.53). The solid lines predict the noise figure with (4.45), while the dashed lines predict the noise figure with (4.53). The agreement is very good near the point of optimum Q_s, thus validating the approximation that $\rho \ll 1$. Figure 4.8 also illustrates the noise figure prediction when the induced gate noise is ignored. Without the gate noise, the model severely underestimates the noise figure.

3.3 COMPARISON WITH THE CLASSICAL APPROACH

The classical approach to minimizing noise figure is presented in detail in Appendix B. This technique, first outlined by Haus et al., approaches the problem by assuming that the device geometry and bias conditions are specified *a priori*, and proceeds to optimize the source impedance to achieve the minimum noise figure under this constraint.

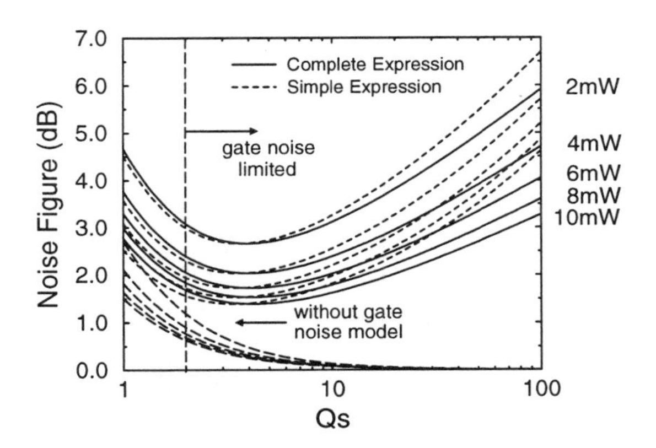

Figure 4.8. Theoretical predictions of noise figure F for several power dissipations. $L = 0.44\mu m$, $R_s = 50\Omega$, $\omega_0 = 10$Grps,$V_{dd} = 2.5$V, $\gamma = 1.3$ [2], $\delta = 2.6$, $|c| = 0.395$ [1], $\nu_{sat} = 1 \times 10^5$ m/s, and $\varepsilon_{sat} = 4.7 \times 10^6$ V/m [3].

For comparison purposes, we repeat the result obtained in the analysis of Appendix B, which is that

$$Q_{s,opt,C} = \frac{\sqrt{\frac{5\gamma}{\delta\alpha^2}} + |c|}{\sqrt{1-|c|^2}} \approx 2.162 \tag{4.58}$$

$$\begin{aligned} F_{min,C} &= 1 + \sqrt{\frac{4}{5}\delta\gamma}\left(\frac{\omega_0}{\omega_T}\right)\sqrt{1-|c|^2} \\ &= 1 + 0.820\sqrt{\delta\gamma}\left(\frac{\omega_0}{\omega_T}\right) \geq 1 + 0.773\left(\frac{\omega_0}{\omega_T}\right). \end{aligned} \tag{4.59}$$

It is interesting to note that the fixed-G_m analysis results in an optimum Q that is very similar to the classical approach. Indeed, the principal difference between these two techniques is that the amplifier is operated off-resonance in the classical solution, due to the fact that $Y_c \neq sC_{gs}$. As proven by Haus et al. [43], the minimum noise figure for any linear twoport is achieved with a particular source conductance when the source susceptance cancels the noise correlation susceptance of the network. Such a condition is commonly referred to as a *conjugate noise match.* A MOS device with partially correlated gate noise has a correlation susceptance given by

$$B_{cor} = sC_{gs}\left[1 + |c|\sqrt{\frac{\delta\alpha^2}{5\gamma}}\right] \leq 1.25sC_{gs}. \tag{4.60}$$

Hence, the optimum source susceptance is an inductance which resonates with the gate capacitance at a frequency slightly *higher* than ω_0. This is sufficient to specify the imaginary part of Y_s. A simple transformation can be used to put the source admittance into a series impedance form which is equivalent at a particular frequency. This transformation preserves the value of inductance for moderate values of Q, thus ensuring that the series resonance will occur at nearly the same frequency as its parallel counterpart. This series equivalent corresponds to the architecture of the LNA.

Because the analysis presented in this book assumes a series resonance at the frequency of operation, we may conclude that it does not quite yield F_{min} for a particular device. However, the difference in the optimum series resonance frequency and ω_0 is only about 15 percent, which explains the similarity between the fixed-G_m optimization and the classical optimization. Hence, we can expect the proposed architecture to possess near-optimum noise performance.

The fixed-P_D analysis, on the other hand, suggests an optimum Q that is quite different from the classical result. This difference compels us to compare the two approaches to determine which one to follow. As we will demonstrate, the fixed-P_D approach is likely to be preferred in most cases.

The difference between these two techniques is one of constraints. The power-constrained approach identifies the best MOS *device* for a specified R_s and power consumption. In contrast, the classical technique seeks to determine the optimum Z_s for a *given* MOS device at a specified power level, and though this latter approach achieves the best noise performance for a particular device, it may yield a sub-optimal result for other figures-of-merit (such as input reflection coefficient).

Additionally, the classical technique reflects the bias of a discrete circuit design standpoint by assuming that the device is fixed while only the input matching network is under the designer's control. In contrast, in the context of integrated circuits, the designer may wish to fix the input matching network and tailor the device geometry to optimize performance. Furthermore, the power-constrained approach permits power consumption to be considered as an explicit parameter, which is useful in low-power systems where this is often an important design constraint.

It is important to emphasize that, although a comparison of (4.57) and (4.59) seems to suggest that the classical approach always yields superior noise performance, these expressions are not directly comparable because ω_T may be different in the two cases. This observation is relevant because the higher Q of the power-constrained result leads to a narrower optimum device with higher current density for a given power consumption. In fact, it is relatively easy to calculate the ω_T ratio for the two cases.

$$\frac{\omega_{T,p}}{\omega_{T,c}} = \frac{g_{m,p}}{g_{m,c}} \frac{C_{gs,c}}{C_{gs,p}} \approx \left(\frac{Q_{s,opt,P_D}}{Q_{s,opt,C}}\right)^{3/2} = 2.28. \tag{4.61}$$

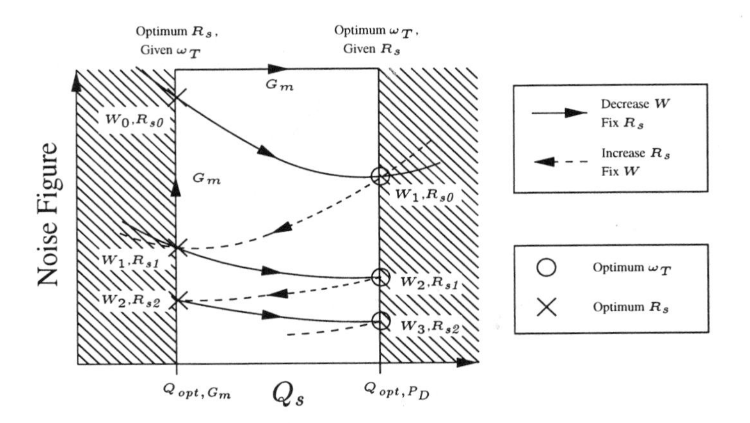

Figure 4.9. Noise figure optimization experiment illustrating the significance of Q_{opt,G_m} and Q_{opt,P_d}. Note that the curves shown represent constant-P_D.

In this expression, the subscript p refers to power-constrained optimization, and c refers to the classical optimization. The resulting minimum noise figures can be directly compared as follows:

$$\frac{F_{min,P_D} - 1}{F_{min} - 1} = \frac{1.426}{0.820} \left(\frac{\omega_{T,c}}{\omega_{T,p}} \right) = 0.763. \tag{4.62}$$

Hence, for a given power consumption, the power-constrained optimization yields superior noise performance and also provides an impedance match, which is a desirable design goal in many cases.

To clarify this point further, consider Figure 4.9, which illustrates the LNA design space for a constant power consumption. In this figure, the solid arcs represent fixed-P_D optimizations, while the dashed arcs represent optimizations where R_s is modified to noise-match the given device. Suppose that we begin with a device which has been optimized using the fixed-G_m analysis for a particular R_{s0}, resulting in a device width W_0. Although R_s is nearly optimally matched to this particular device, superior noise performance can be obtained at the same power dissipation by decreasing the device width to W_1, following the fixed-P_D arc. The noise performance improves in this procedure despite the non-optimal source resistance because ω_T improves as the scaling is performed. This increase offsets the loss in noise match until Q_{opt,P_D} is reached. At this point, the gate noise dominates the output noise of the device. So, degrading the noise match in favor of the gate noise permits operation at an elevated ω_T; the net result is improved noise performance. Also note that the gain, G_m, actually *improves* in this procedure.

Of course, once the new width, W_1, is determined, an increased source resistance can be found which is noise-matched to this new device. This procedure takes the design back along the dashed arc, yielding improved noise performance until Q_{opt,G_m} is reached. However, there is a significant penalty in G_m which is incurred by this increase in R_s (recall that G_m is inversely proportional to R_s). Nonetheless, this procedure could be repeated (at the expense of G_m) as long as it is reasonable to increase R_s and decrease W, maintaining Q_s to lie within the white region of Figure 4.9.

The question is: at what point (and at which Q_s) should the ultimate design be placed? Assuming that a maximum realistic R_s can be specified, it seems reasonable always to design the LNA to operate at Q_{opt,P_D} because this design point will always possess a larger G_m than its lower-Q counterpart. The result is that the best LNA design operates at a Q_s which is *different* from the value corresponding to the conjugate noise match. A noise mismatch is tolerated in return for a higher ω_T at the same power dissipation.

We conclude that the optimization procedures given here, though not yielding F_{min} precisely as outlined in [43], permit selection of the best device for *two* constraints simultaneously: perfect input match, and a specific gain; or perfect input match, and a specific power dissipation. Of these, the second set of constraints yields the best combination of noise, power, and gain. There is only *one* device in a given technology that optimizes noise performance while satisfying either set of two of these specifications for a particular R_s.

Finally, it is clear that, in all cases, the minimum noise figure improves as ω_T increases with advances in technology. This fact, taken in conjunction with the experimental results of this study, signifies that CMOS low noise amplifiers will soon achieve noise performance at GPS frequencies that is largely parasitic-limited, making CMOS an attractive alternative to more costly silicon bipolar and GaAs technologies.

3.4 A NOTE ON MOS NOISE SIMULATION MODELS

The preceding analysis facilitates the design of CMOS low-noise amplifiers using this topology. It is important to note, however, that existing MOS noise models — as implemented in circuit simulators such as HSPICE — do not adequately account for hot-electron effects or induced gate effects. The options available for level 13, 28, and 39 MOS models (BSIM-I, Modified BSIM-I, and BSIM-II, respectively) do not account for even the most elementary of short-channel noise effects, much less the more advanced considerations of the previous section. This situation is particularly disturbing, given that the optimal LNA design will undoubtedly be limited by the gate noise of the device.

Some strides have been made recently with the adoption of the BSIM-III model. This model makes use of an alternative formulation for channel thermal noise in which the noise power is treated as proportional to the total inversion layer charge [88]. This is the same model proposed by *Wang et al.* [62]. Short-channel effects can be included in the formulation of the inversion layer

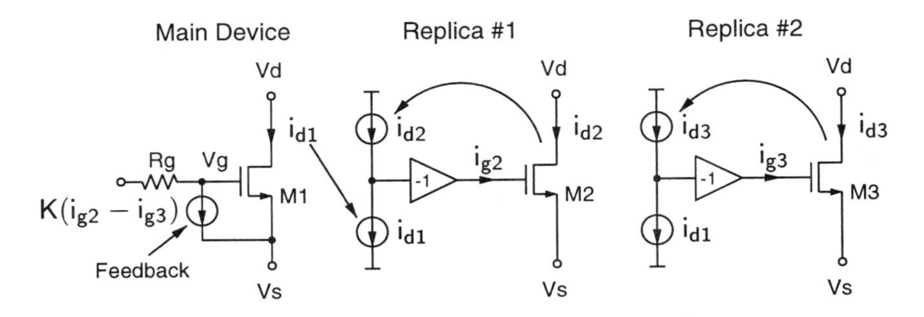

Figure 4.10. Modified NMOS noise model that includes the effects of induced gate noise and gate polysilicon resistance.

charge, and hence in the noise power. However, even this model discounts the possibility that elevated carrier temperature is an important factor. The assumption of a uniform carrier temperature along the entire channel length may explain the departure of the model's predictions from measured data for relatively short-channel devices [89].

To circumvent the absence of a gate noise model in modern CMOS device simulators, there are two courses of action. The first course is to modify the device model equations to include the gate noise term. This modification is possible in some simulation environments where the simulator code is accessible to the designer. In such cases, the gate noise can easily be included if it is represented as a noisy *charge* fluctuation that is added to the total gate charge in the model. In this case,

$$\frac{\overline{Q_g^2}}{\Delta f} = 4kT\delta\frac{C_{gs}^2}{5g_{d0}}. \tag{4.63}$$

This is the approach that was taken for the GPS receiver implementation described in Chapter 7.

A second technique employs a macromodel in Spice to generate a gate noise current with the right power spectral density that roughly tracks variations in g_m so that the designer is free to consider device geometry tradeoffs. Figure 4.10 illustrates this technique, which is applicable to any MOS simulator. Device M1 is the device that we wish to model. Its drain current is measured and reproduced in two replica devices, M2 and M3, through the use of current feedback loops, implemented in Spice by current-controlled current sources. By replicating the terminal voltages of M1 in biasing M2 and M3, we can ensure that all three devices conduct precisely the same current. Thus, the resulting gate voltages of M2 and M3 are exactly equal to V_g, except for one thing. Because M1, M2 and M3 generate thermal noise, the fed-back gate voltages are noisy. In fact, the gate voltage of M2 will have a noise power of precisely

$$\overline{v_{g2}^2} = \frac{8kTB\gamma}{g_m} \tag{4.64}$$

assuming, for simplicity, that $g_m \approx g_{d0}$. The reason for the factor of 8 instead of the usual 4 is that both M1 and M2 contribute equally to the noise voltage. Note that, mathematically, this voltage has the same expression as twice the input-referred squared noise voltage of the device. This voltage causes a current to flow in the gate of the device. The current, i_{g2}, is given by

$$\overline{i_{g2}^2} = \frac{8kTB\gamma\left[\omega\left(C_{gs} + C_{gd}\right)\right]^2}{g_m}. \tag{4.65}$$

Now, because half of this current is due to the drain noise of M1, these two sources are partially correlated. The same is true of i_{g3}, so that if we subtract i_{g3} from i_{g2}, the correlated parts cancel, leaving a noise current that has no correlation with the drain current of M1. This current is fed back to the gate of M1 as $K\left(i_{g2} - i_{g3}\right)$. The constant, K, is selected by comparing the expression for the induced gate noise

$$\overline{i_g^2} = \frac{4kTB\delta\left(\omega C_{gs}\right)^2}{5g_m} \tag{4.66}$$

with equation (4.65). If we assume that $\delta \approx 2\gamma$, then we can solve for K.

$$K = \frac{1}{\sqrt{5}\left(1 + \frac{C_{gd}}{C_{gs}}\right)}. \tag{4.67}$$

With this choice for K, we have a model that produces an equivalent gate noise with the right magnitude and frequency dependence that properly tracks variations in device geometry and bias. In addition, by subtracting i_{g3} from i_{g2}, any signal currents due to M1 cancel as well, thus ensuring that the feedback does not interfere with the operation of the device.

3.5 ADDITIONAL DESIGN CONSIDERATIONS

The noise performance of a complete LNA is fundamentally limited by the noise figure of the amplifier input stage, as described in detail in the preceding sections. However, optimizing the input stage is a necessary, but insufficient step to guarantee optimal noise performance for the whole amplifier. This section describes a number of techniques that will help to maximize LNA performance and avoid unnecessary and costly design oversights. Additional detail is available in Appendix C which presents experimental results for both a single-ended and a differential CMOS LNA.

The preceding sections provide a thorough analysis of the common-source LNA input stage, which affords the best possible noise performance of any architecture. In this analysis, the effect of the gate-drain overlap capacitance C_{gd} was neglected for simplicity. However, the influence of C_{gd} cannot be neglected in practice. From simulations of CMOS LNA circuits, including the effect of the induced gate noise, it has been found that the power-constrained optimum Q_s remains approximately constant for realistic values of C_{gd}, assuming that a cascode input stage is used to reduce the Miller effect. Based on this observation, a useful rule-of-thumb is that the optimum device width is inversely proportional to frequency. That is,

$$W_{opt} \approx \frac{\kappa}{f_0} \tag{4.68}$$

where $\kappa = 500\mu$m-GHz, based on simulations in a 0.5-μm technology. Note that because the capacitance per unit gate width is relatively constant as technologies scale, this value for κ should remain approximately the same for shorter channel lengths.

In addition to selecting the correct device width, it is important to consider whether to use a single-ended or differential design. Single-ended amplifiers will consume half the power and die area for a given theoretical noise performance when compared to differential amplifiers. Furthermore, differential amplifiers typically require an off-chip balun that introduces additional loss, thereby degrading the overall noise figure. On the other hand, single-ended amplifiers are much more sensitive to ground inductance and substrate impedances than are differential amplifiers. In particular, the common substrate inductance can severely compromise reverse isolation and even amplifier stability, as demostrated by the experimental results in Appendix C. Furthermore, a single-ended LNA will be more susceptible to substrate and supply noise, which is an important issue for single-chip receivers. In contrast, differential amplifiers possess a degree of common-mode rejection, and the substrate impedance is of secondary concern. These advantages of the differential approach greatly simplify the design of a complete receiver system.

A number of additional principles that constitute good LNA design practice can be enumerated. In particular,

- The use of a cascode input stage leads to improved gain and improved amplifier stability by eliminating interaction between the input matching circuit and the output tuned circuit. It is important, however, to minimize the noise contribution of the cascode device by minimizing the capacitance at its source. Merging the source of the cascode device with the drain of the input device proves to be an effective technique for reducing this capacitance.

- When selecting L_s to generate a 50Ω real term in the input impedance, it is important to ensure that one does not inadvertently use too little inductance. Doing so will lead to a severe noise penalty, as shown in Appendix C.
- The input devices should be laid out with multiple gate fingers to reduce the total polysilicon gate resistance, and the device should be surrounded with substrate contacts to reduce the effective back-gate resistance.
- Although it is tempting to perform the input matching on-chip with simple inductive tuning, the resistive losses associated with spiral inductors are typically prohibitively large for this purpose because of the noise figure degradation that they introduce.
- One should be careful to shield the LNA input pads from the substrate to eliminate unnecessary loss. In addition, the use of patterned ground shields beneath the spiral inductors greatly reduces the energy lost to the substrate [52].

Clearly, a number of important considerations play a direct role in determining the ultimate LNA performance that is achieved. Fortunately, none of them represents a fundamental barrier and all can be mitigated with sufficient attention to detail in the design and layout of the amplifier.

4. SUMMARY

We have presented a thorough analysis of the design of CMOS low noise amplifiers in this chapter. With a corrected MOS noise model in hand, the power-constrained optimization procedure leads to simple design criteria that yield the best combination of noise figure, input power match and gain.

Theoretical analysis of the amplifier architecture has demonstrated the fundamental role of induced gate noise, which is essential in defining the minimum noise figure. That in many practical cases this source of noise may *dominate* the output noise of the amplifier underscores the critical need for improved MOS noise models. Given the intense interest in RF CMOS, it is likely that improved models will be developed in the near future.

Experimentally, as detailed in Appendix C, we have demonstrated a single-ended 30mW low noise amplifier in a 0.6-μm CMOS process that is suitable as a first amplifier in a GPS receiver. The amplifier's noise figure of 3.5dB was the lowest reported for a CMOS LNA in this frequency range when it was first published. A second experimental LNA was presented that achieves a 3.8dB noise figure with only 12mW of power consumption in a 0.35-μm CMOS process. This result is most directly comparable to a single-ended design that consumes 6mW, and thus represents a substantial improvement over the first LNA. The low power consumption of this architecture is a result of aggressive current conservation through current-reuse in the input and output stages.

In the final implementation of the GPS receiver in Chapter 7. we will demonstrate a differential LNA that achieves a 2.4-dB noise figure with 12-mW power consumption in a 0.5-μm CMOS process. This result represents a significant advance in CMOS low noise amplifiers, and is directly enabled by the theoretical basis presented in this chapter.

Chapter 5

CMOS MIXERS

An essential element in all modern radio receivers is the mixer, which is responsible for frequency translation or "heterodyning". Historically, a family of techniques have developed for performing frequency translation, including the double-balanced diode-ring mixer and the Gilbert multiplier, among others. More recently, research has been focused on extending these techniques for use in CMOS technologies and on developing new techniques that are uniquely suited for CMOS.

This chapter begins in section 1. with a review of commutating mixer techniques. In section 2., we present a double-balanced CMOS voltage mixer that has been selected for use in this work. A detailed analysis of the mixer reveals some unexpected conversion gain behaviors that arise due to the linear time-variant (LTV) nature of the mixer when its output is capacitively loaded. In addition, we will address the topics of noise figure and linearity to illustrate the merits of this architecture.

1. REVIEW OF MIXER ARCHITECTURES

The predominant class of mixers that find use in modern radio receivers is the time-varying or "commutating" class of mixers. Unlike nonlinear mixers that employ cross-modulation to perform mixing, commutating mixers employ switching mechanisms to change periodically the sign of the input signal under the control of a local oscillator. Because the RF and IF ports are linearly related (at least in principle), such mixers generally exhibit improved distortion characteristics compared to their nonlinear counterparts.

Commutation can be performed in the voltage or current domains. Figure 5.1 illustrates three mixer examples; two that use a diode ring for voltage switching, and one that uses bipolar current switches. In Figure 5.1(a), the local oscillator is coupled to the center taps of the two transformers to control

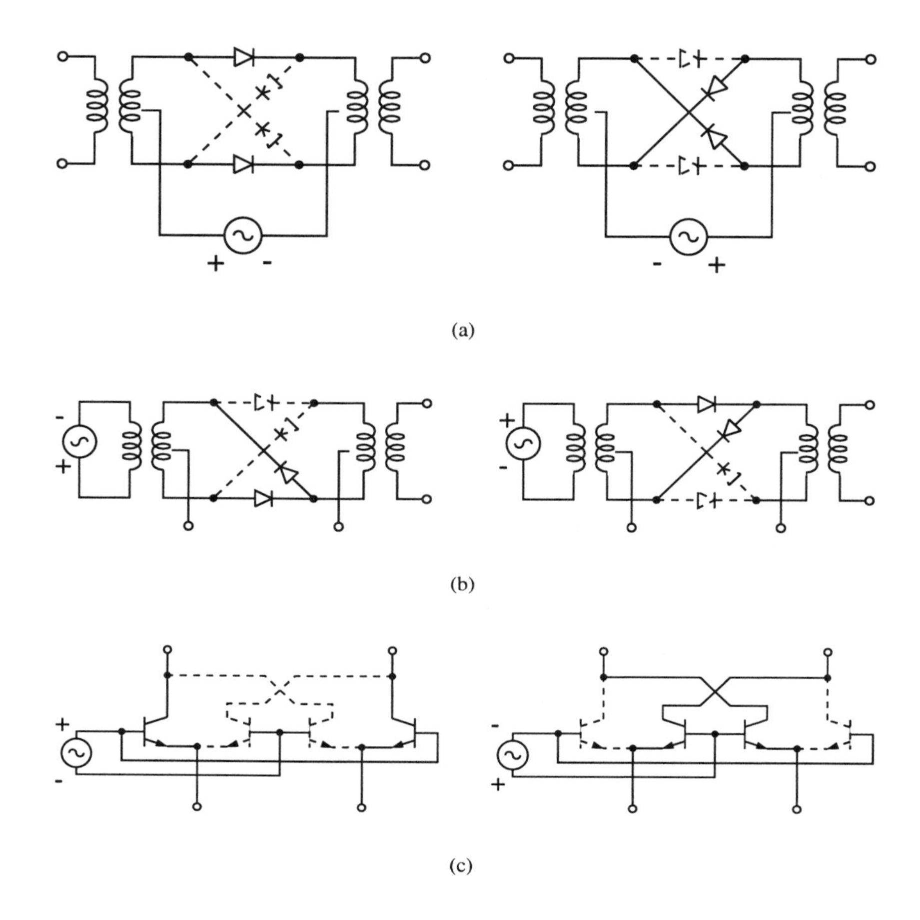

Figure 5.1. Commutating mixer architectures, illustrating the switching principle employed in each. (a) Diode ring with center-tapped LO drive. (b) Diode ring with transformer-coupled LO drive. (c) Gilbert mixer.

which pair of diodes is forward biased. When the outer pair is forward biased, the input and output ports are connected in phase; when the inner pair is forward biased, the ports are connected out of phase. Because the LO is coupled to the center taps, it does not couple to either of the other two ports. In addition, the symmetric switching prevents input signal frequencies from appearing directly at the output. Because of these isolation properties, the diode ring mixer is called a double-balanced mixer.

Figure 5.1(b) illustrates a different mode of operation for the diode ring mixer in which the LO drive is transformer-coupled to the ring. In this case, a different pair of diodes is activated on each phase of the LO that couples the

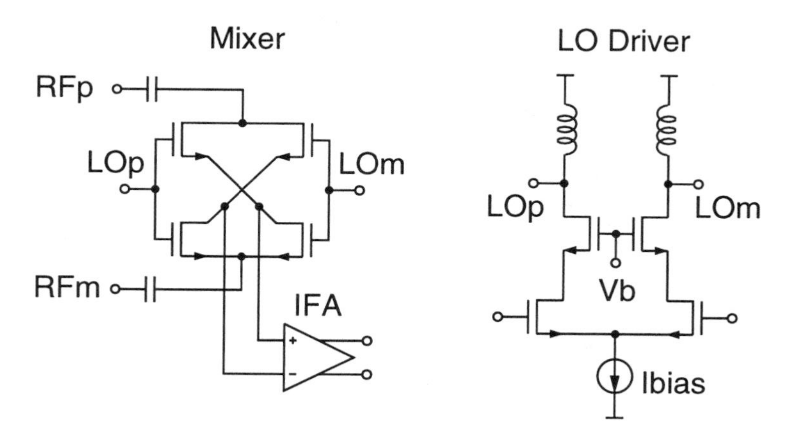

Figure 5.2. CMOS voltage mixer and LO driver.

other transformer to the center taps of the two secondaries. This connection is widely used when the RF and LO signals are at high frequencies and the IF output is at a relatively low frequency. Center tapping the IF port permits the use of smaller self inductances in the transformer secondaries, which only need to present large impedances at the LO and RF frequencies.

While diode ring mixers operate on a voltage switching principle, the Gilbert mixer operates on a current switching principle. As shown in Figure 5.1(c), the mixer employs two bipolar current switches whose outputs are connected in opposition. On one phase of the LO, the input current flows to the output through the outer pair of devices, while on the opposite phase, the input current is diverted to the opposite outputs via the inner pair of devices. If the bases of the transistors are driven symmetrically, the emitters lie at points of symmetry so that, in principle, no local oscillator signal couples to the input or output ports. So, just as in the diode ring case, the Gilbert mixer is nominally a double-balanced mixer.

The Gilbert mixer owes its success in part to the fact that bipolar devices make excellent current switches. By analogy, MOS devices are excellent voltage switches, and this observation motivates an investigation of the CMOS voltage mixer presented in the next section.

2. THE DOUBLE-BALANCED CMOS VOLTAGE MIXER

A mixer topology consisting of a pair of differential voltage switches is illustrated in Figure 5.2. The gates of the switches are driven by a tuned LO driver whose output inductive loads resonate with the gate capacitance of

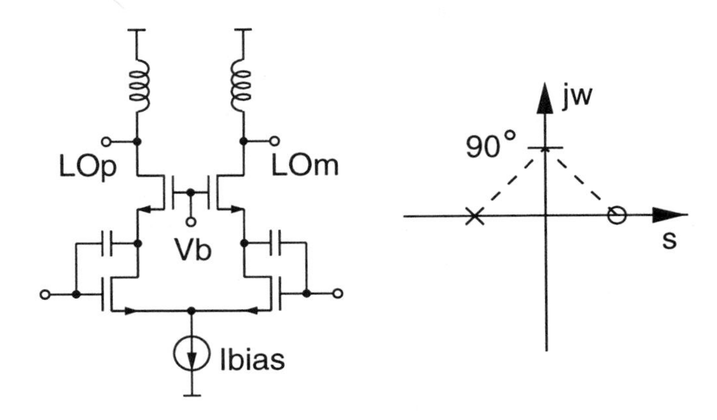

Figure 5.3. Quadrature generation with the Miller capacitance.

the switches to provide a high impedance at the resonance frequency, thereby reducing the power consumption in the driver.

In this mixer, two switches are controlled by the positive phase of the local oscillator, and the other two are controlled by the negative phase, which lags by 180°. Thus, the mixer connects its RF port to its IF port through these switches with a polarity that alternates at the LO frequency. It is this alternation of polarity that establishes mixing.

The receiver uses two of these mixers, driven by quadrature local oscillators. To generate the quadrature LO, one of the LO drivers is modified as shown in Figure 5.3. With the Miller feedback capacitance, the amplifier has a transconductance of

$$G_m = g_m \frac{1 - sC/g_m}{1 + sC/g_m}. \tag{5.1}$$

This expression has a LHP pole and a RHP zero, resulting in an all-pass characteristic with a phase shift of 90° at $\omega = g_m/C$. To regulate the pole/zero frequency, a constant-g_m bias source generates I_{bias}. This technique is suitable for use in this particular receiver due to the relaxed requirements for I/Q matching.

In the following sections, we examine in detail the conversion gain properties, noise figure and linearity of this mixer architecture.

2.1 BASIC MIXER CONVERSION GAIN

The conversion gain of the mixer can be determined by a careful analysis with special attention paid to the time-varying nature of the mixer. There are several types of LO drive that might be applied to the switches in the mixer, as shown in Figure 5.4.

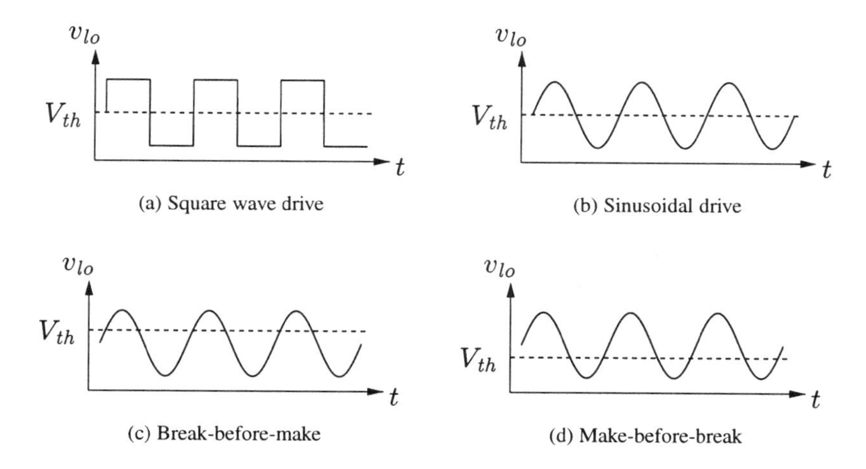

Figure 5.4. Four LO signals investigated

The simplest of these is a square wave with 50% duty cycle, illustrated in Figure 5.4(a). The LO voltage for a square wave drive can be expressed mathematically as follows:

$$v_{lo}(t) = A_{LO}\Pi(2t/T_{LO}) * \sum_{n=-\infty}^{\infty} \delta(t - nT_{LO}) \tag{5.2}$$

where A_{LO} is the local oscillator amplitude, $\Pi(t)$ is the rectangle function, and T_{LO} is the local oscillator period. This LO signal will serve as a reference for comparison with other types of LO signals.

In practice, a square wave drive is difficult to achieve at radio frequencies. A more practical and power efficient method is to resonate the gate capacitances and drive the gates sinusoidally, as shown in Figures 5.4(b)–5.4(d). In this case,

$$v_{lo}(t) = A_{LO}\, cos(2\pi f_{LO}t + \phi_{LO}) + B_{LO} \tag{5.3}$$

where B_{LO} is the DC level on the gates. The choice of B_{LO} determines whether or not the two switch pairs will conduct during overlapping portions of the LO period. In Figure 5.4(b), B_{LO} equals the switch threshold voltage, V_{th}; in 5.4(c), $B_{LO} < V_{th}$, resulting in break-before-make switching action, while in 5.4(d), $B_{LO} > V_{th}$, resulting in make-before-break action.

The switches in the mixer are simply time varying conductances, as shown in Figure 5.5(a). Therefore, we can replace the switch network with a Thévenin equivalent network, as shown in Figure 5.5(b). The open circuit voltage is then

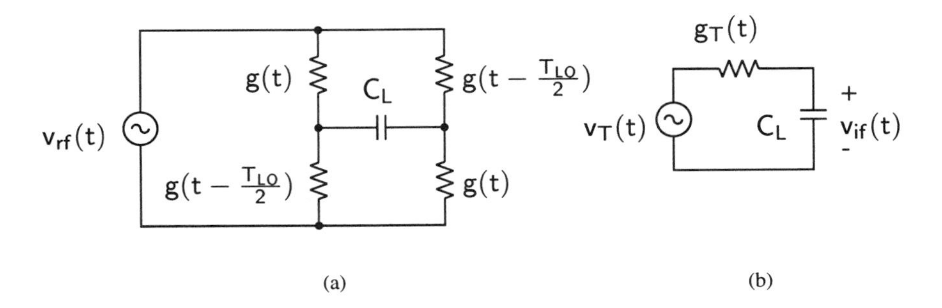

Figure 5.5. Mixer core. (a) Time-varying conductance model, and (b) Thévenin equivalent circuit.

given by

$$v_T(t) = \frac{g(t) - g(t - T_{LO}/2)}{g(t) + g(t - T_{LO}/2)} v_{rf}(t) = m(t) v_{rf}(t) \tag{5.4}$$

and the Thévenin impedance, written as a conductance, is

$$g_T(t) = \frac{g(t) + g(t - T_{LO}/2)}{2}. \tag{5.5}$$

In equation (5.4), $m(t)$ represents the effective *mixing function* of the network. For example, when the LO drive is a square wave, $m(t)$ is a square wave with zero DC value and unit amplitude. This multiplies the input voltage, $v_{rf}(t)$, to yield a mixed open circuit voltage, $v_T(t)$. Figure 5.6 illustrates the mixing function:

$$m(t) = \frac{g(t) - g(t - T_{LO}/2)}{g(t) + g(t - T_{LO}/2)} \tag{5.6}$$

and $g_T(t)$, for the four types of LO drive presented in Figure 5.4.

Both $m(t)$ and $g_T(t)$ exhibit important properties. The mixing function $m(t)$ has no DC component, is periodic with a period of T_{LO}, and has half wave symmetry, implying that it only has odd frequency content (nf_{LO}, where n is an odd integer). The absence of a DC component in the mixing function indicates that the RF voltage is isolated from the IF port of the mixer. In contrast, the conductance $g_T(t)$ has a DC component and is periodic with period $T_{LO}/2$. The average conductance plays an important role in setting the IF bandwidth of the mixer, as will be shown shortly.

If we assume that $C_L = 0$, then $v_{if}(t) = v_T(t)$. To find the conversion gain from the RF port to the IF port, we evaluate the magnitude of the Fourier transform of the mixing function at f_{LO}. The resulting conversion gain, G_c, is

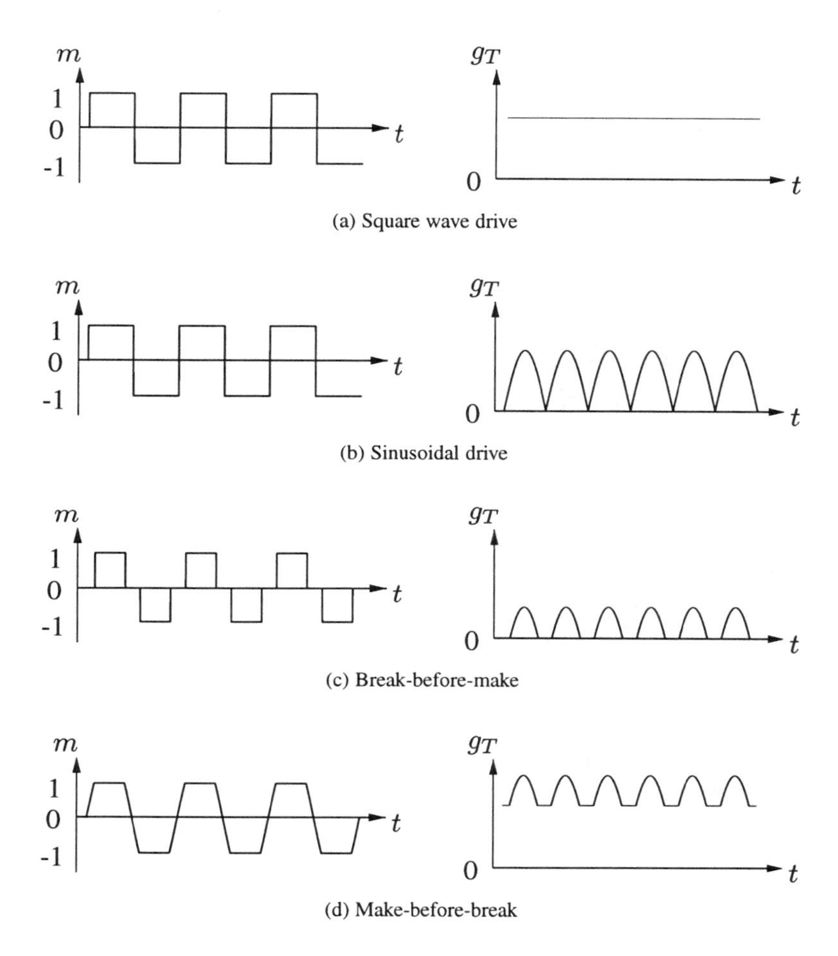

Figure 5.6. Mixing function and Thévenin conductance for the four cases

shown in Table 5.1 for the four types of LO drive. It is interesting to note that G_c in the last two cases involves a single parameter, r, that characterizes the choice of B_{LO} and A_{LO}.

2.2 LTV CONVERSION GAIN ANALYSIS

In general, C_L does not equal zero. To solve for the conversion gain in this case, we must determine the time-varying impulse response of the network and apply the superposition integral to find $v_{if}(t)$ as a function of $v_{rf}(t)$. In doing so, we will demonstrate that capacitive loading can be used to actually *increase* the conversion gain of the mixer above the classical limit of -3.92dB.

Table 5.1. G_c for the four types of LO drive.

Square wave drive	$2/\pi$	
Sinusoidal drive	$2/\pi$	
Break-before-make	$(2/\pi)\sqrt{1-r^2}$	
Make-before-break	$\frac{sin^{-1}(r)/r+\sqrt{1-r^2}}{\pi}$	$0 \le r \le 1$
	$1/(2r)$	$1 \le r < \infty$
	$r = \frac{\lvert V_{th}-B_{LO}\rvert}{A_{LO}}$	

To evaluate the impulse response, we apply an impulse to the circuit in Figure 5.5(b) at time τ, $v_T(t) = \delta(t-\tau)$. The initial voltage produced on C_L can most readily be determined by transforming the Thévenin equivalent circuit in Figure 5.5(b) into its Norton form with the following short circuit current:

$$i_N(t) = g_T(t)v_T(t) = g_T(\tau)\delta(t-\tau). \tag{5.7}$$

The total charge delivered to the capacitor as a result of this impulse in current is $g_T(\tau)$ coulombs. This charge produces an initial voltage of $g_T(\tau)/C_L$ volts on C_L at time τ. The following differential equation describes the circuit's response to this initial condition:

$$C_L\frac{dv_{if}(t)}{dt} = -g_T(t)v_{if}(t). \tag{5.8}$$

The solution has the form $h(t) = Ae^{-f(t)}$. Combining the initial condition with this solution, and noting that the system is causal, yields the network impulse response:

$$h(t,\tau) = \frac{g_T(\tau)}{C_L}e^{-\int_\tau^t \frac{g_T(s)}{C_L}ds}u(t-\tau) \tag{5.9}$$

where $u(t)$ is the unit step function. To determine the response at the IF port of the mixer, we simply apply the superposition integral for this impulse response with an input voltage of $m(\tau)v_{rf}(\tau)$. The result is that

$$v_{if}(t) = \int_{-\infty}^{t} \frac{g_T(\tau)}{C_L}e^{-\int_\tau^t \frac{g_T(s)}{C_L}ds}m(\tau)v_{rf}(\tau)d\tau. \tag{5.10}$$

To simplify this expression, it is useful to express $g_T(t)$ as a Fourier series

$$g_T(t) = \overline{g_T} + \sum_{n=1}^{\infty} a_n cos(n2\omega_{LO}t + \phi_n) = \overline{g_T} + \widetilde{g_T}(t) \tag{5.11}$$

where $\overline{g_T}$ is the DC level of $g_T(t)$ and $\widetilde{g_T}(t)$ is the time-varying portion of $g_T(t)$. Furthermore, we define the integral of $\widetilde{g_T}(t)/C_L$ to be

$$\widetilde{f_T}(t) = \frac{\overline{g_T}}{2\omega_{LO}C_L} \sum_{n=1}^{\infty} \frac{a_n sin(n2\omega_{LO}t + \phi_n)}{n\overline{g_T}} + K \tag{5.12}$$

where K is an arbitrary constant. With these definitions, we can re-express the superposition integral in (5.10) as follows:

$$v_{if}(t) = e^{\widetilde{f_T}(t)} \int_{-\infty}^{t} \frac{\overline{g_T}}{C_L} e^{-\frac{\overline{g_T}}{C_L}(t-\tau)} e^{-\widetilde{f_T}(\tau)} \frac{g_T(\tau)}{\overline{g_T}} m(\tau) v_{rf}(\tau) d\tau. \tag{5.13}$$

Equation (5.13) can be simplified by paying attention to the form of $\widetilde{f_T}$. Note that from equation (5.12), $\widetilde{f_T}$ is proportional to a normalizing coefficient

$$\frac{\overline{g_T}}{2\omega_{LO}C_L} \tag{5.14}$$

which is the ratio of the average bandwidth of the network to twice the local oscillator frequency. Thus, when C_L is sufficiently large that the average bandwidth is much less than the local oscillator frequency, the exponential terms involving $\widetilde{f_T}$ reduce to unity, and (5.13) can be simplified, yielding

$$v_{if}(t) = \int_{-\infty}^{t} \frac{\overline{g_T}}{C_L} e^{-\frac{\overline{g_T}}{C_L}(t-\tau)} \frac{g_T(\tau)}{\overline{g_T}} m(\tau) v_{rf}(\tau) d\tau. \tag{5.15}$$

We can identify the function of each term in (5.15) to clarify the influence of the load capacitance, C_L on the mixer's operation. First, the RF port voltage is multiplied by a modified mixing function, which is conveniently defined to be

$$m'(t) = \frac{g_T(t)}{g_{Tmax}} m(t). \tag{5.16}$$

In this expression, g_{Tmax} is the peak conductance of $g_T(t)$, which normalizes $m'(t)$ to vary between ± 1. This modified mixing function appears in Figure 5.7 for the four LO drive signals analyzed previously. Second, by introducing g_{Tmax}, we can also define an effective gain

$$A = \frac{g_{Tmax}}{\overline{g_T}} \tag{5.17}$$

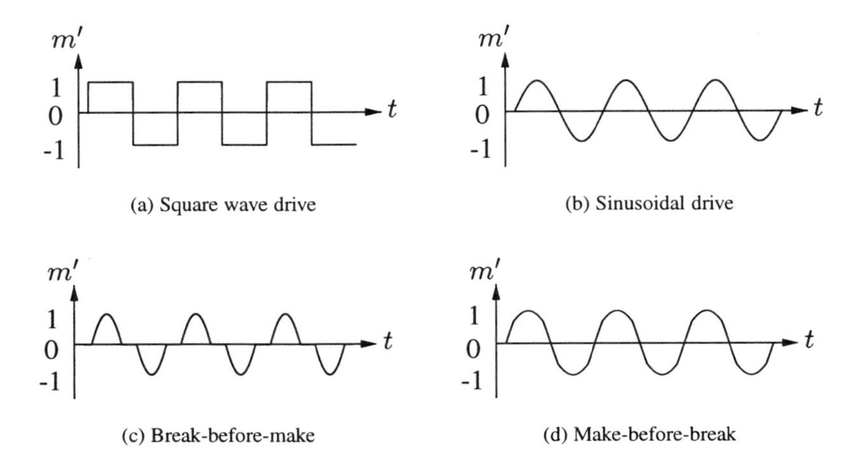

Figure 5.7. Modified mixing functions for the four cases

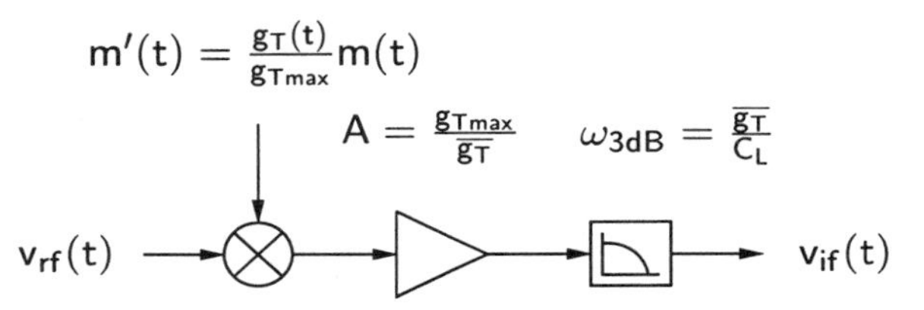

Figure 5.8. Equivalent block diagram for core conversion gain

which is the ratio of the peak conductance to the average conductance. Finally, the remaining terms simply describe a lowpass filter with bandwidth $\overline{g_T}/C_L$. Combining these concepts, (5.15) can be expressed as

$$v_{if}(t) = h_{lpf}(t) * \left[Am'(t)v_{rf}(t)\right] . \tag{5.18}$$

An equivalent block diagram representation of this expression appears in Figure 5.8. The total voltage conversion gain is just $AG_c'|H_{lpf}(f_{IF})|$, where G_c' is the conversion gain associated with $m'(t)$, and $|H_{lpf}(f_{IF})|$ is the gain of the effective low-pass filter at the IF frequency.

It is interesting to compare the conversion gain for a sinusoidal LO drive of the capacitively terminated mixer to that of the reference square wave drive. For a sinusoidal drive, $G_c' = 1/2$, whereas for a square wave drive, $G_c' = 2/\pi$. But in the square wave case, the peak-to-average conductance is unity, while in the sinusoidal case it is $\pi/2$. Thus, the voltage conversion gain for a sinusoidal drive with capacitive loading is actually $\pi/4$ (-2.1dB), which *exceeds* the $2/\pi$ (-3.9dB) value for a square wave drive.

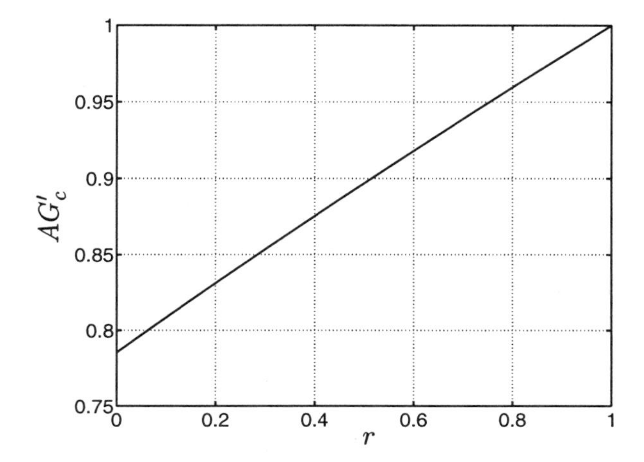

Figure 5.9. AG'_c vs. r for a break-before-make LO drive.

The analysis can be taken one step further by considering the conversion gain for a break-before-make drive. It is possible, though slightly involved, to express conversion gain as

$$\begin{aligned} AG'_c =& \frac{cos^{-1}(r) - r\sqrt{1-r^2}}{2[\sqrt{1-r^2} - r\,cos^{-1}(r)]} \\ \approx& \left(1 - \frac{\pi}{4}\right) r + \frac{\pi}{4} \end{aligned} \tag{5.19}$$

where $r = (V_{th} - B_{LO})/A_{LO}$. Figure 5.9 plots (5.19) as a function of r. Surprisingly, the voltage conversion gain actually approaches unity as $r \rightarrow 1$, which corresponds to the extreme condition of the break-before-make drive where each switch is on for a single instant of the LO cycle. Thus, in principle, by adjusting the switching point of the mixer switches, the conversion loss can be made arbitrarily small. However, as r approaches unity, linearity suffers greatly, making this mode of operation impractical.

As shown in this section, the conversion gain for a capacitively loaded CMOS voltage mixer can be increased by using a sinusoidal LO drive. This effect is a direct result of the time-varying filter formed by the switch on-resistance and the load capacitance. In the GPS receiver of this work, this effect is exploited to improve the conversion gain of the mixers by using the input capacitance of the IF amplifier as the terminating capacitance.

2.3 MIXER NOISE FIGURE

Although the mixer core consumes negligible power, the LO driver consumes power to drive the input capacitance of the mixer. Thus, the sizing of the mixer devices is linked to the design of the LO driver. The size of these devices and the LO drive level also determine the noise figure of the mixer. So, the mixer noise figure and LO driver power consumption are intimately linked.

As demonstrated in Chapter 2, the SSB noise figure of an ideal mixer (one that has no internal resistive losses) is given by

$$F = L_c \tag{5.20}$$

where L_c is the power conversion loss of the mixer. This equation can be modified to include the on-resistance of the mixer switches, which appears in series with the source resistance. Thus,

$$F = L_c \left[1 + \frac{2\overline{R_m}}{R_s}\right] \tag{5.21}$$

where $\overline{R_m}$ is the average on-resistance of a single switch.

To determine $\overline{R_m}$, we can assume that

$$\overline{R_m} = \frac{1}{g_{ds}} = \frac{L}{\mu_{eff} C_{ox} W \overline{V_{od}}} \tag{5.22}$$

where $\overline{V_{od}}$ is the average overdrive voltage supplied by the LO driver. This voltage is related to the power consumption in the LO driver, the frequency of operation and the Q of the spiral inductors that are used. The choice of spiral inductor is also constrained by the requirement that it resonate with the input capacitance of the mixer plus other parasitic capacitances, including the spiral's own self-capacitance.

If the LO driver has a tail current of I_0, the amplitude of one of the LO driver outputs is given by

$$A_{LO} = \frac{\eta_I I_0 Q_L}{2\omega_{LO} C} \tag{5.23}$$

where Q_L is the Q of the load inductors, and C is the total load capacitance on the LO driver output. The parameter η_I accounts for the fact that not all of the bias current is available at the output, due to parasitic losses in the LO driver itself. The capacitance, C, comprises two switch gate capacitances plus the spiral's self-capacitance. If the driver has a certain self-resonant frequency, ω_{SR}, then the allowed switch capacitance is given by

$$C_{ox} W L = \frac{C}{2}\left[1 - \frac{\omega_{LO}^2}{\omega_{SR}^2}\right]. \tag{5.24}$$

Finally, combining equations (5.21) through (5.24), we find an approximate formula for the mixer noise figure:

$$F = L_c \left[1 + \frac{\omega_{LO}/\omega_C}{1 - \left(\omega_{LO}/\omega_{SR}\right)^2}\right] \tag{5.25}$$

where ω_{SR} is the self-resonant frequency of the spiral inductors, and ω_C is a critical frequency given by

$$\omega_C = \frac{R_s Q_L \mu_{eff} \eta_I I_0}{8\pi L^2} \tag{5.26}$$

beyond which the achievable noise figure begins to degrade rapidly. For a reasonable choice of self-resonant frequency, the noise figure is degraded by 3dB when $\omega_{LO} = \omega_C$.

As a brief example, suppose that $\omega_{LO} = 10$Grps, $\omega_{SR} = 30$Grps, $R_s = 100\Omega$, Q_L=5, $\mu_{eff} = 250\text{cm}^2/\text{Vs}$, $\eta_I = 0.75$, $I_0 = 4\text{mA}$ and $L = 0.5\mu\text{m}$. Then $\omega_C \approx 6$Grps, and $F = 2.9L_c$, resulting in a loss of 4.6dB over the ideal (internally lossless) mixer. If R_s is increased to 400Ω, then $F = 1.5L_c$, which is a loss of 1.7dB when compared to the ideal case.

Although this analysis is greatly simplified, the result yields some intuition about fundamental tradeoffs. In particular, the performance of this type of mixer should improve dramatically as technology scales due to the $1/L^2$ factor in ω_C. For a given bias current, the noise performance is strongly influenced by the Q_L of the spiral inductor loads. In addition, the self-resonant frequency should be no lower than about $3\omega_{LO}$ to avoid a sharp degradation in noise figure.

Notably absent from the result is any dependence on the switch size. As one reduces the switch size (and increases the load inductance value), the voltage swing at the gate of the switch increases so that the average on resistance, $\overline{R_m}$, remains roughly constant. So, the size of the switch is a relatively free parameter that can be optimized for maximum linearity.

There is one additional merit of the CMOS voltage mixer structure that is worth mentioning. Because there is no DC current through the switches, they contribute no $1/f$ noise to the mixer output. This consideration is particularly important in direct conversion architectures where low-frequency noise and offsets at the mixer output are of concern.

For a more thorough treatment of the general subject of noise in mixers, the interested reader can refer to [90].

2.4 MIXER LINEARITY

There are two major sources of distortion in the mixer: device nonlinearities and phase modulation of the switching instants.

To improve the linearity of the transistors, it is most important to keep the current through the switches small to reduce nonlinear voltage drops across the devices [91]. This criterion is satisfied with the use of a small capacitive load, which presents a high impedance to the output. Note, however, that this requirement is at odds with the use of capacitive loading to boost the conversion gain of the mixer. The remaining nonlinearities consist of parasitic junction capacitances, which are weak nonlinearities.

A second source of distortion arises from phase modulation of the mixing function by the RF voltage, an effect which is also found in diode-ring modulators [92]. This distortion becomes more pronounced as the amplitude of the RF voltage approaches that of the LO drive voltage. In this situation, the instant at which switching occurs exhibits a significant dependence on the RF voltage itself. Hence, the effective mixing function depends directly on the RF voltage and this dependence introduces distortion. Borrowing from the research on diode ring mixers, we may expect this type of distortion to diminish if larger LO drive levels are used to steepen the LO waveform's slope as it passes through zero. A corollary is that square wave drives will typically lead to improved linearity over sinusoidal drives. References [91] and [92] contain more detailed treatments of this type of distortion in diode ring mixers.

3. SUMMARY

This chapter has presented a thorough analysis of the double-balanced CMOS voltage mixer, which uses four CMOS voltage switches to implement a mixer with no static power consumption. When reactively loaded, this mixer exhibits some unusual conversion gain properties that are only predictable by treating the mixer as a bandwidth-limited linear time-variant (LTV) system. In particular, the conversion gain can exceed the classical limit of $2/\pi$ when the mixer is capacitively loaded.

In addition, we have presented a noise analysis of the mixer that demonstrates its potential for low-noise operation. Combined with the fact that a large voltage headroom is available due to the simplicity of the mixer topology, this architecture offers the promise of wide dynamic range, as long as the LO drive is sufficiently strong.

The mixer is the last block in the signal path that processes RF signals. As we move on to analyze the IF chain, we enter the world of baseband amplifier techniques. The need for large "inductances" in the frequency-selective blocks operating at these lower frequencies forms the principal motivation for the considerations of the next chapter. In particular, we will explore how to maximize the dynamic range of one of the most vexing subsystems in an integrated receiver: the on-chip active filter.

Chapter 6

POWER-EFFICIENT ACTIVE FILTERS

The channel filter of an integrated receiver is arguably one of the most important signal path blocks, and one of the most difficult to realize in integrated form. It is responsible for attenuation of out-of-band interference that might de-sensitize the receiver, and thus it must possess a large stop-band rejection. In integrated form, the realization of high-Q filters is complicated by the lack of suitable integrated inductors. Although active circuit techniques can provide the necessary reactance through feedback, the relatively large noise and low dynamic range of active inductors limits their use substantially.

In this chapter, we begin by reviewing some techniques for implementing integrated lowpass filters. Then, in section 2., we pursue a thorough analysis of one of these techniques, the G_m-C technique, in which active inductors (or gyrators) are used to obtain complex poles and zeros. The limitations of this approach are made explicit by the analysis, and design guidelines are elucidated. In particular, a figure of merit for active transconductor architectures is derived that permits a fair comparison of different approaches. Section 3. presents a survey of transconductor architectures based on this figure of merit, arriving finally at a new transconductor that is both power-efficient and linear.

1. PASSIVE AND ACTIVE FILTER TECHNIQUES

The passive L-C ladder filter that enjoys widespread use today was invented in 1915 nearly simultaneously by Wagner in Germany and Campbell in the U.S. [32] [33]. It appears that Wagner had priority, but his work was suppressed by German military authorities during World War I. The basic filter topology invented by Wagner is shown in Figure 6.1. In its original incarnation, the "electric wave-filter" comprised a chain of identical, repeating sections. However, the filter theory was soon extended to allow for sections of differing

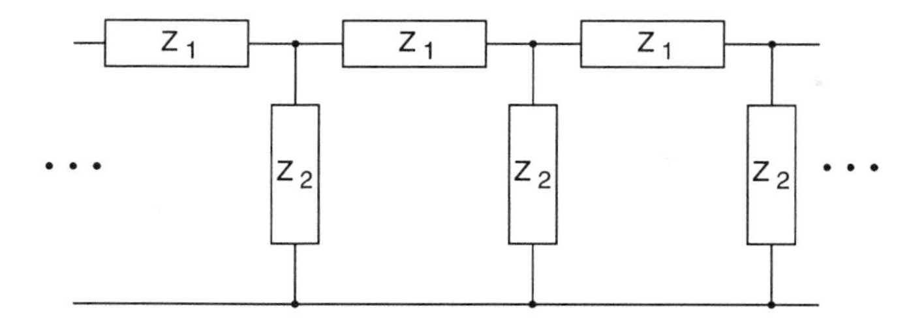

Figure 6.1. Generic structure of the "electric wave-filter", or ladder filter.

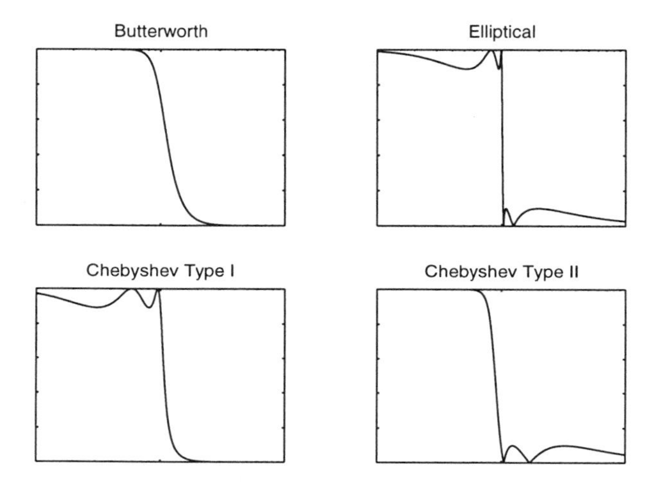

Figure 6.2. Four common types of lowpass filters.

impedances [34]. Later theoretical developments enabled the realization of a variety of frequency and phase responses.

Figure 6.2 illustrates several common lowpass filter types that are realizable with a finite number of ladder sections. The Butterworth filter has a maximally flat passband, while the other filtershapes provide sharper cutoffs with equal ripple either in the passband (Chebyshev Type I), the stopband (Chebyshev Type II), or both (elliptical). For the present work, the elliptical filter is significant because it possesses the sharpest transition band of any filter topology for a given order and is therefore the most selective of all the filters. It also has the most severe phase distortion. However, for the GPS system, the phase distortion is not of primary concern due to the large processing gain.

Passive ladder filters can be implemented using discrete inductors and capacitors, by exploiting mechanical resonance in quartz crystals, or by using

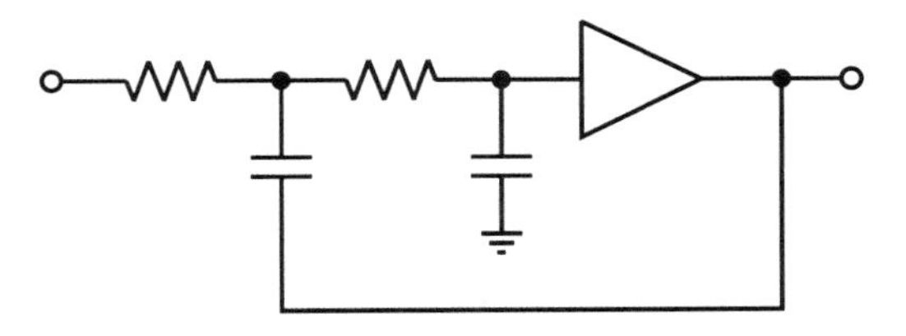

Figure 6.3. A Sallen and Key lowpass filter.

acoustic waves in ceramic materials. In integrated receivers, one can also employ passive spiral inductors and capacitors, but the low quality factor and small inductance of integrated inductors severely limits their utility [93]. Recently, there has been some interest in micromechanical filters in silicon that operate at frequencies as high as 70MHz [94]. Using micromachined resonators, one can achieve very high quality factors. However, these filters presently suffer from a number of important problems that remain to be solved. These include poor component tolerances, tiny dynamic range, and the need to operate in a near vacuum to realize reasonable quality factors. Until these problems are resolved, micromechanical filters are not an attractive alternative to off-chip passive filters.

Due to the lack of suitable passive components, integrated filters that operate in the low-MHz range of frequencies must be implemented using active circuit techniques. There are, however, several tradeoffs that must be considered when determining whether or not to integrate a filter. Integrated active filters consume power, produce distortion and noise, and consume die area. These factors place integrated filters at an immediate disadvantage compared to off-chip discrete filters. All the same, there are a few good reasons for integrating the filters in a radio receiver. One reason for using integrated filters is the reduction of board cost and component count that accompanies their use. A second reason is the reduction in electromagnetic interference that may occur due to board level coupling from other integrated circuits. Finally, in the case of GPS receivers, complete system integration helps to enable embedded applications where GPS functionality can easily be integrated into more complex systems, such as cellular phones. For these reasons, integration of the IF filters has been pursued in this work.

To realize integrated active filters, there are a number of techniques to consider. One family of filters, known as Sallen and Key filters [95], uses a network of resistors and capacitors and a single feedback amplifier. Because RC networks only produce real poles, feedback is required to realize complex poles. An example of a Sallen and Key approach is shown in Figure 6.3. This

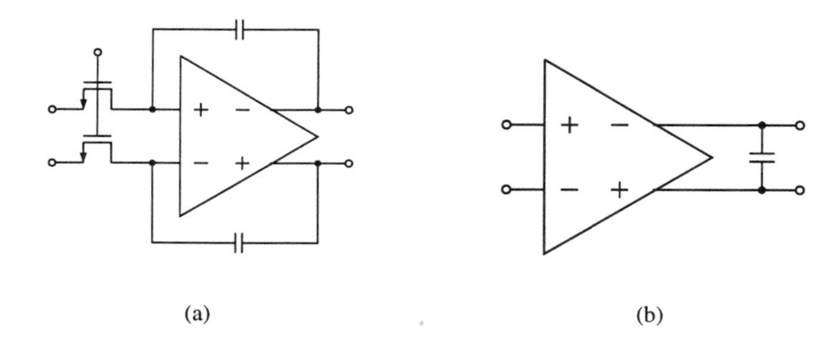

Figure 6.4. Integrators for the (a) MOSFET-C filter and (b) G_m-C filter.

circuit implements a second order transfer function of the form

$$H(s) = \frac{h}{s^2 + ds + 1}. \tag{6.1}$$

Although filters of this sort are very simple, their principal limitation is an extreme sensitivity to process variations [96].

There are other active filters that use integrators as fundamental building blocks. Complex filter responses can be obtained by enclosing several integrators in various types of feedback loops. Filters that use this approach include MOSFET-C filters and G_m-C filters. The basic integrator element of each filter type is illustrated in Figure 6.4.

In the MOSFET-C integrator, a simple operational amplifier integrator is used with two FET's acting as variable resistors. Tuning is accomplished by adjusting the gate voltages of the two FET's. In contrast, the G_m-C integrator uses a transconductor driving a load capacitance. Tuning is then accomplished by setting the transconductance. There are various methods for automatically tuning these filters that have been extensively studied in the literature. A discussion of tuning techniques is beyond the scope of this work, but the interested reader can consult [97] for numerous examples.

In the present work, which emphasizes low power for portability, the MOSFET-C technique stands at a disadvantage due to the need for operational amplifiers. Though typically less linear, G_m-C filters offer reduced power consumption due to their efficient use of supply current. Because of this, the G_m-C technique was selected for this work.

The following sections examine the limitations of G_m-C filters in some detail to determine how to maximize dynamic range for a given power consumption. In particular, we will look at a subclass of G_m-C filters called *gyrator* filters that directly implement the inductive elements of an LC ladder using gyrator circuits of the type shown in Figure 6.6.

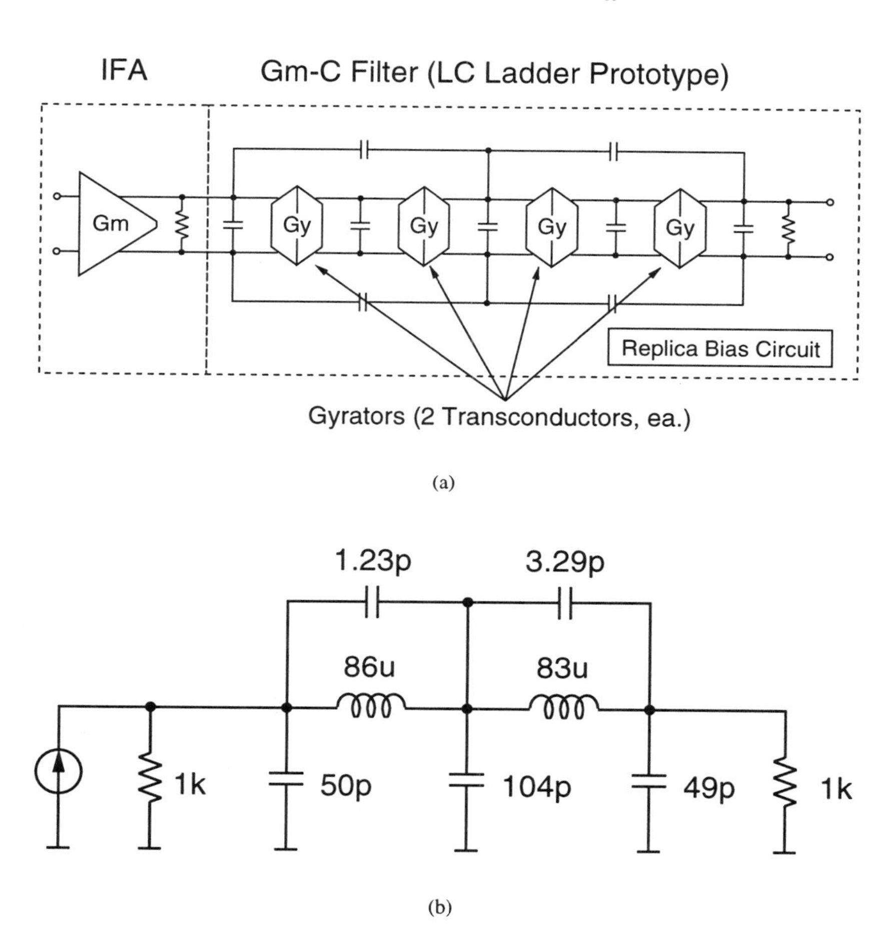

Figure 6.5. Block diagram of the on-chip G_m-C filter and its equivalent half-circuit.

2. DYNAMIC RANGE OF THE ACTIVE G_M-C FILTER

Figure 6.5 illustrates a block diagram of the on-chip active filter and its equivalent half-circuit. The intermediate-frequency amplifier (IFA) drives the input of the filter directly, and the load resistors in the IFA output stage also terminate the filter input. Similarly, a real resistor provides the output termination, permitting a reduction in power consumption.

As shown in Table 3.1, the filter is the dynamic range limiting block in the system. Previous studies have typically examined the dynamic range problem in active filters by assuming that the largest acceptable signal voltage is a fixed parameter [98], often expressed as a simple fraction of the supply voltage

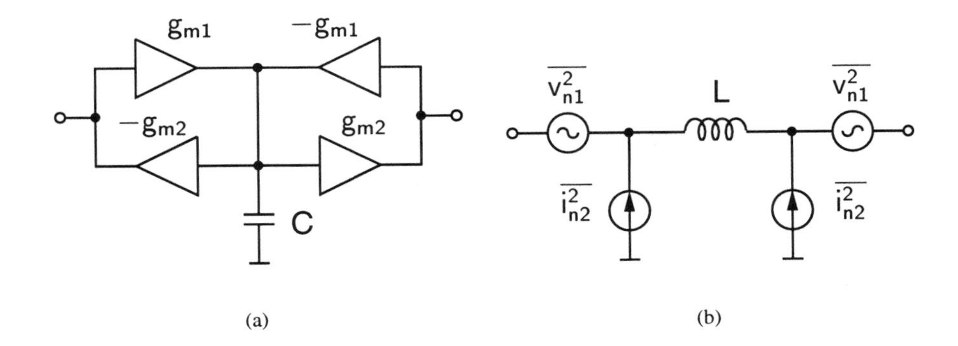

Figure 6.6. A simple gyrator and its equivalent circuit with noise sources.

[53]. One problem with such an approach is that it partly obscures the dependence of dynamic range on power consumption and choice of transconductor architecture. In previous work where the dynamic range is formulated explicitly without such assumptions, the analysis is typically limited to the specific architecture under discussion, and is therefore lacking in generality [99] [100].

In the following discussion, we derive an expression for dynamic range with power consumption as an explicit constraint that is broadly applicable to G_m-C filters, independent of transconductor architecture. In doing so, we will identify a transconductor figure of merit that can be applied to aid the selection of an architecture that maximizes the dynamic range of the filter for a given power consumption.

We begin by deriving the minimum noise figure of the filter.

2.1 NOISE FIGURE

To determine the noise figure, we construct a noise model of each gyrator to understand how its internal amplifiers contribute noise to the system.

Figure 6.6(a) shows an equivalent circuit of a simple gyrator that implements a floating inductor. Each transconductor generates thermal noise at its output that degrades the noise figure of the filter. By referring the transconductor noise sources to the external terminals of the gyrator, one arrives at the equivalent circuit shown in Figure 6.6(b), where

$$L = \frac{C}{g_{m1} g_{m2}} \tag{6.2}$$

$$\frac{\overline{v_{n1}^2}}{\Delta f} = \frac{4kT\epsilon}{g_{m1}} \tag{6.3}$$

$$\frac{\overline{i_{n2}^2}}{\Delta f} = 4kT\epsilon g_{m2} \tag{6.4}$$

where $\epsilon > 1$ is a factor describing the amount of excess noise generated by a transconductor cell when compared to a real conductance of the same value.

At low frequencies, the inductor presents a short, and the voltage and current noise sources contributed by all of the gyrators in the filter sum together so that

$$\overline{v_n^2} = 2N_L\overline{v_{n1}^2} \tag{6.5}$$

$$\overline{i_n^2} = 2N_L\overline{i_{n2}^2} \tag{6.6}$$

where N_L is the number of inductors in the filter. The corresponding minimum spot noise figure is

$$F_{min} = 2\left[1 + 2\epsilon N_L\sqrt{\frac{g_{m2}}{g_{m1}}}\right] \approx 4\epsilon N_L\sqrt{\frac{g_{m2}}{g_{m1}}} \tag{6.7}$$

which occurs for an optimum terminating resistance of

$$R_t = \frac{1}{\sqrt{g_{m1}g_{m2}}}. \tag{6.8}$$

It is interesting to note that the minimum noise figure depends primarily on the order of the filter (through N_L) and on the architecture of the transconductor (through ϵ). In fact, the minimum noise figure does not depend on the choice of R_t, implying that a fixed *power* gain is required preceding the filter to minimize its contribution to the system noise figure. Some improvement can be obtained by adjusting the relative magnitudes of g_{m1} and g_{m2}, but this degree of freedom is constrained by the need for good distortion performance, as shown in the next section.

2.2 3RD-ORDER INTERMODULATION DISTORTION

The analysis of distortion mechanisms in the filter is considerably more complex than the noise analysis of the previous section. In a communications system, however, certain simplifications can be made by restricting the analysis to 3rd-order intermodulation (IM3) distortion and ignoring the more difficult case of harmonic distortion. The virtue of IM3 distortion for analytical purposes is that the distortion products lie close to the fundamental products as long as the fundamental tones are close to one another. In what follows, we will assume that the fundamental frequencies are *arbitrarily* close to one another.

To begin, we adopt the assumption of small levels of distortion. With this assumption, we can model the distortion of a given transconductor in one of

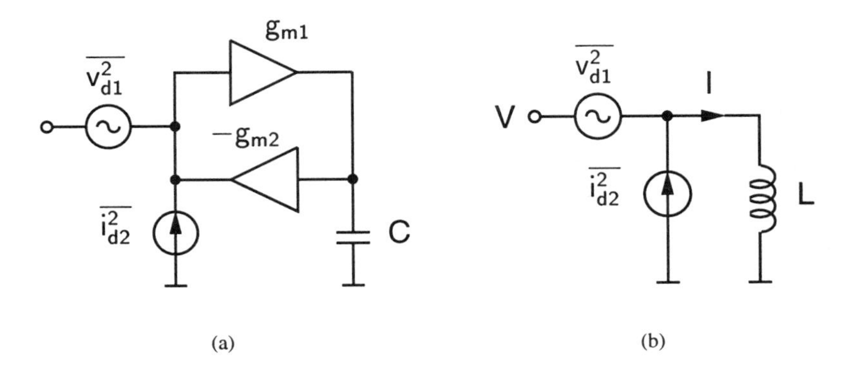

Figure 6.7. Distortion models for a gyrator. (a) Full gyrator. (b) Equivalent circuit.

two ways. In the first method, we replace the nonlinear transconductor with a linear one and attribute the distortion products to an additive current source in the output of the transconductor. Alternatively, we can refer the output distortion current to the input as an equivalent input-referred distortion voltage. The distortion current and voltage have magnitudes given by

$$|v_d| = |V| \left(\frac{|V|}{V_{IP3}} \right)^2 \tag{6.9}$$

$$|i_d| = |I| \left(\frac{|I|}{I_{IP3}} \right)^2 \tag{6.10}$$

where

$$I_{IP3} = g_m V_{IP3} \tag{6.11}$$

is a measure of the third-order intercept point of the transconductor. The voltage and current intercept points are related by g_m because they are extrapolated from low-amplitude distortion measurements, below the onset of gain compression.

Using both of these transconductor distortion models in a simple gyrator results in the circuit shown in Figure 6.7. This construction illustrates that it is important to consider both the voltage swing across the gyrator input g_{m1} as well as the current swing through the feedback transconductor g_{m2}.

The greatest distortion will occur when the largest signal voltage appears across the gyrator. This condition corresponds to the resonance of the inductor with the other filter elements. Because the inductor voltage and current are in quadrature with one another, and because the fundamental frequencies are

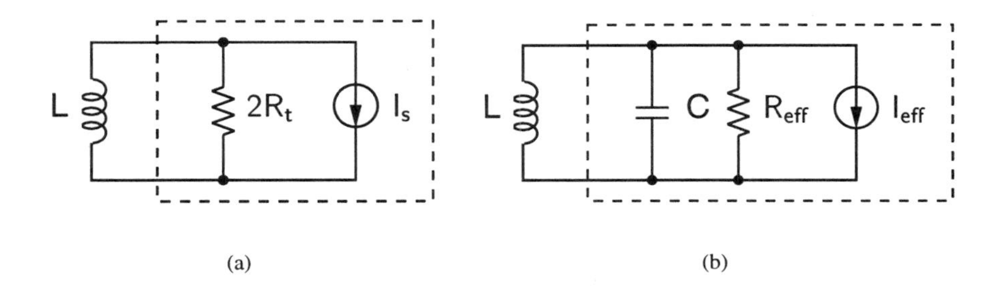

Figure 6.8. Equivalent circuit presented by the filter network to the inductor (a) at dc, and (b) at resonance.

arbitrarily close to each other, we can assume that the resulting distortion products are also in quadrature. Thus, the total distortion voltage appearing across the inductor at resonance is

$$|v_{dt}|^2 = |v_{d1}|^2 + (\omega_0 L Q_r)^2 |i_{d2}|^2 \tag{6.12}$$

where Q_r is the Q of the resonance.

Noting that $|V| = \omega_0 L|I|$, and combining (6.8)–(6.12), we can express the total distortion voltage as

$$|v_{dt}|^2 = \frac{|V|^6}{V_{IP3}^4}\left[1 + \frac{Q_r^2}{Q_t^4}\left(\frac{g_{m1}}{g_{m2}}\right)^2\right] \tag{6.13}$$

where we have defined

$$Q_t = \frac{\omega_0 L}{R_t} = \omega_0 L\sqrt{g_{m1} g_{m2}}. \tag{6.14}$$

Finally, to relate the magnitude of the inductor voltage, $|V|$, to the source voltage, V_s, consider Figure 6.8, which illustrates the equivalent circuits presented by the filter to the inductor at dc and at resonance. The filter acts as an impedance transforming network, causing the termination resistance $2R_t$ to be transformed to an effective parallel resistance R_{eff} at resonance. The source current I_s is also transformed to an effective current I_{eff} by the inverse square root of the impedance transformation ratio. Thus,

$$\frac{|V|}{V_s} = \frac{I_{eff}}{I_s}\frac{R_{eff}}{2R_t} = \sqrt{\frac{R_{eff}}{2R_t}} \tag{6.15}$$

or, in terms of Q_r and Q_t,

$$\frac{|V|}{V_s} = \sqrt{\frac{Q_r Q_t}{2}}. \tag{6.16}$$

By substituting (6.16) into equation (6.13), we can relate the distortion voltage to the source voltage.

$$|v_{dt}|^2 = \frac{|V_s|^6}{V_{IP3}^4} \frac{Q_r^3 Q_t^3}{8} \left[1 + \frac{Q_r^2}{Q_t^4}\left(\frac{g_{m1}}{g_{m2}}\right)^2\right] = \frac{|V_s|^6}{V_{IP3,eff}^4} \tag{6.17}$$

where $V_{IP3,eff}$ is the effective IIP3 voltage, referred to the input of the filter.

This analysis has, so far, assumed that only one inductor is present. With multiple inductors, the analysis becomes more complex. However, we can form a pessimistic bound by assuming that the N_L inductors contribute equal amounts of distortion power. With this assumption, the IIP3 *available power* is

$$\text{IIP3} = \frac{V_{IP3,eff}^2}{8R_t N_L^{\frac{1}{2}}} \approx \frac{V_{IP3}^2}{R_t N_L^{\frac{1}{2}} (2Q_r Q_t)^{\frac{3}{2}} \left[1 + \frac{Q_r^2}{Q_t^4}\left(\frac{g_{m1}}{g_{m2}}\right)^2\right]^{\frac{1}{2}}}. \tag{6.18}$$

Although this expression is approximate, it yields some insight on how the IIP3 will depend on the relative magnitudes of various parameters. In particular, N_L, Q_r and Q_t are set by the desired filter characteristic, and are thus relatively inflexible parameters. Also, note that reducing the impedance level of the filter will result in improved distortion because voltage levels are reduced for a given signal power. Finally, the ratio of g_{m1} and g_{m2} strongly influences the linearity because this ratio determines the current-handling capability of the active inductor.

In the next section we will explore how to optimize the dynamic range of the filter, based on these results for noise figure and IIP3.

2.3 OPTIMIZING DYNAMIC RANGE

The condition for minimum noise figure expressed in equation (6.8) determines the *product* of g_{m1} and g_{m2}. The *ratio* of these transconductances is a free parameter that can be used to maximize dynamic range. The peak spurious-free dynamic range (SFDR) is the dynamic range for which IM3 distortion products and the in-band noise power are equal. In terms of F and IIP3, we have

$$SFDR = \left[\frac{IIP3}{FktB}\right]^{2/3}. \tag{6.19}$$

Using (6.7) and (6.18), we can formulate the SFDR as

$$SFDR = \left[\frac{V_{IP3}^2}{4kTB\epsilon\,(2N_L Q_r Q_t)^{\frac{3}{2}}\, R_t \left[\frac{g_{m2}}{g_{m1}} + \frac{Q_r^2}{Q_t^4}\frac{g_{m1}}{g_{m2}}\right]^{\frac{1}{2}}}\right]^{\frac{2}{3}}. \tag{6.20}$$

Maximizing this expression is equivalent to minimizing

$$R_t \left[\frac{g_{m2}}{g_{m1}} + \frac{Q_r^2}{Q_t^4} \frac{g_{m1}}{g_{m2}} \right]^{\frac{1}{2}}. \tag{6.21}$$

The condition for minimum noise figure is expressed in equation (6.8). Substituting this for R_t in (6.21) yields

$$\left[\frac{1}{g_{m1}^2} + \frac{1}{g_{m2}^2} \frac{Q_r^2}{Q_t^4} \right]^{\frac{1}{2}}. \tag{6.22}$$

We can minimize (6.22) subject to a constant-power constraint if we set

$$g_{m1} + g_{m2} = \frac{\beta P_D}{2N_L} = \beta \overline{P_D} \tag{6.23}$$

where $\overline{P_D}$ is the power dissipation per gyrator, and β is the transconductance per unit power dissipated in the gyrator. Taking the derivative of (6.22) and setting it equal to zero yields the condition for maximizing dynamic range, which is that

$$\frac{g_{m2}}{g_{m1}} = \frac{Q_r^{2/3}}{Q_t^{4/3}}. \tag{6.24}$$

So, in general, it is *not* optimal to have $g_{m1} = g_{m2}$. This is particularly true of high-Q filters, which tend to have a larger optimum ratio of the two transconductances due to larger circulating currents in the inductors at resonance.

Combining (6.23) and (6.24) with (6.20), we can express the peak $SFDR$ as

$$SFDR_{pk} = \frac{\left[\frac{\overline{P_D}}{4kTB} \frac{\beta V_{IP3}^2}{\epsilon} \right]^{\frac{2}{3}}}{2N_L Q_r Q_t \left[1 + \frac{Q_r^{2/3}}{Q_t^{4/3}} \right]}. \tag{6.25}$$

This expression for dynamic range deserves close attention. The denominator is determined entirely by the *filter* architecture. In particular, the higher the Q and the greater the number of inductors, the lower the dynamic range will be, assuming all other factors are held constant. So, architectures that relax the required filter order and Q will benefit from increased dynamic range. The numerator of the expression is determined by the *transconductor* architecture, including $\overline{P_D}$, the power dissipated per gyrator. Note that expending more power increases the dynamic range by lowering the optimum terminating impedance. The form of the numerator suggests that a good figure of merit for a transconductor architecture is

$$\Gamma_m = \frac{\beta V_{IP3}^2}{\epsilon} \tag{6.26}$$

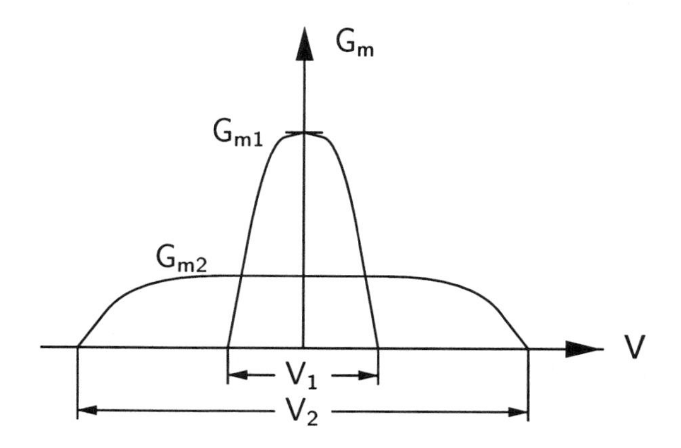

Figure 6.9. Illustration of the class-A G_m for linearity tradeoff.

which is a dimensionless quantity because β has units of V^{-2}. To maximize Γ_m, it is important to select a transconductor architecture that is power-efficient (high β), linear (high V_{IP3}) and low-noise (small ϵ). Choosing such an architecture forms the subject of the next section.

3. POWER-EFFICIENT TRANSCONDUCTORS

The gyrator transconductor architecture sets the overall performance of the filter. To implement a low-power filter, it is essential to select a transconductor architecture that is linear and that maximizes the G_m/I_{bias} ratio, thereby maximizing β.

Class-A techniques are fundamentally limited in their power efficiency, as shown in Figure 6.9. This figure illustrates two possible transconductance curves for a hypothetical class-A transconductor. The input voltage range can be increased at the expense of G_m by using simple feedback, such as source degeneration. However, in both cases the maximum input voltage is limited by

$$G_{m1}V_1 = G_{m2}V_2 \propto I_{bias}. \tag{6.27}$$

In words, the area under the G_m curve is approximately constant. This tradeoff between G_m and linear input voltage range makes class-A techniques unattractive because the only recourse for increasing the linear range for a given G_m is to increase the power consumption.

In contrast, class-AB techniques are more flexible because the bias current increases in the presence of large signal excitations. Thus, the standing power can be smaller while maintaining large-signal linearity.

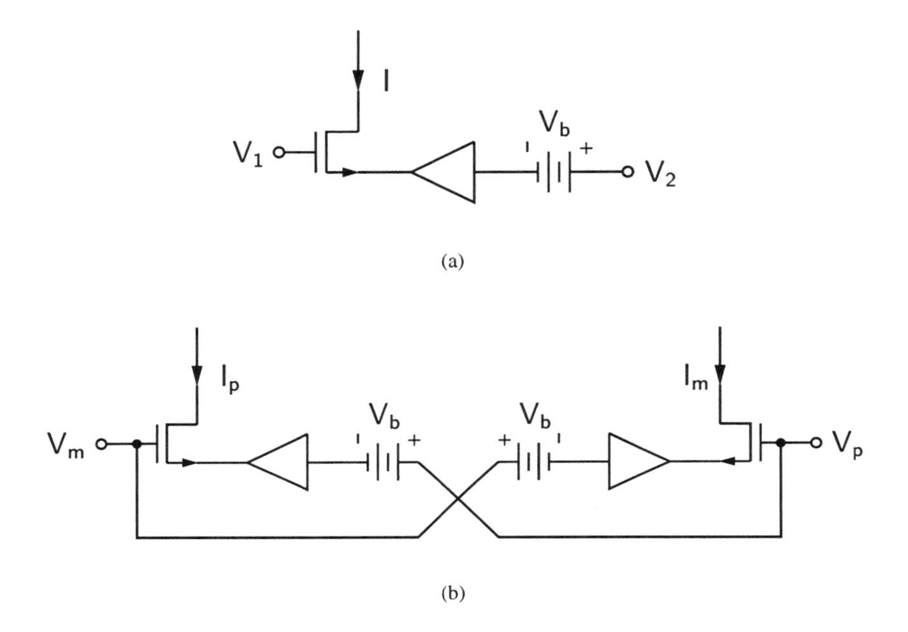

Figure 6.10. A class-AB transconductor. (a) Square-law prototype. (b) Linear prototype.

3.1 A CLASS-AB TRANSCONDUCTOR

One approach to implementing a power-efficient, linear transconductor in a CMOS technology is to take advantage of the square-law behavior of the devices themselves. If a differential amplifier is constructed out of two square-law amplifiers, the resulting differential gain is *linear*. For two such transconductors, with differential input Δv,

$$\beta_0 \left(\Delta v + V_b\right)^2 - \beta_0 \left(-\Delta v + V_b\right)^2 = 4\beta_0 V_b \Delta v. \tag{6.28}$$

Thus, the output is linearly proportional to the input.

To implement a square-law transconductance characteristic, one might consider the circuit of Figure 6.10(a). In this circuit, the input voltage, $V_1 - V_2$, is level-shifted and placed across an NMOS device. To the extent that the NMOS follows a square-law, the overall transconductor is square-law. Using two of these transconductors as shown in Figure 6.10(b), we can construct a linear differential transconductor. However, due to velocity saturation and vertical field mobility degradation, the NMOS will exhibit sub-square-law behavior. Velocity saturation can be mitigated by adopting a longer channel length, but vertical field mobility degradation depends on oxide thickness, which is not a flexible parameter.

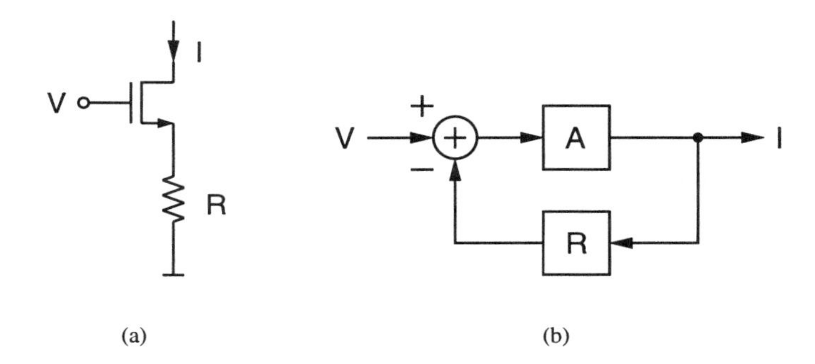

Figure 6.11. Mobility degradation modeled as series feedback. (a) Equivalent circuit. (b) System view.

To understand the role of mobility degradation, one can model this effect with an ideal square-law device and simple series feedback, as shown in Figure 6.11. If the mobility-degraded current is given by

$$I = \frac{\beta_0 (V_{gs} - V_T)^2}{1 + \theta (V_{gs} - V_T)} \tag{6.29}$$

then one may model the degradation as the result of an ideal square-law device degenerated by a resistor of value

$$R = \frac{\theta}{2\beta_0}. \tag{6.30}$$

In the equivalent system model, the resistor appears as a negative feedback term, with the forward path, A, representing the desired squaring operation

$$I = \beta_0 (V)^2. \tag{6.31}$$

When viewed from this perspective, it is clear that one remedy for mobility degradation is positive feedback, as shown in Figure 6.12. The proper amount of positive feedback can be selected by setting k to

$$k \approx g_m R = \frac{g_m \theta}{2\beta_0} = \theta V_{od} \left[\frac{1 + \theta V_{od}/2}{(1 + \theta V_{od})^2} \right] \tag{6.32}$$

where $V_{od} = V_{gs} - V_T$ with the inputs balanced. In reality, k must be adjusted to compensate for second order effects due to other non-idealities, such as channel-length modulation and body effect. Thus, equation (6.32) is only approximate. Note that a practical value of k for this process is about 0.2. As a result, this small amount of positive feedback does not pose a stability threat.

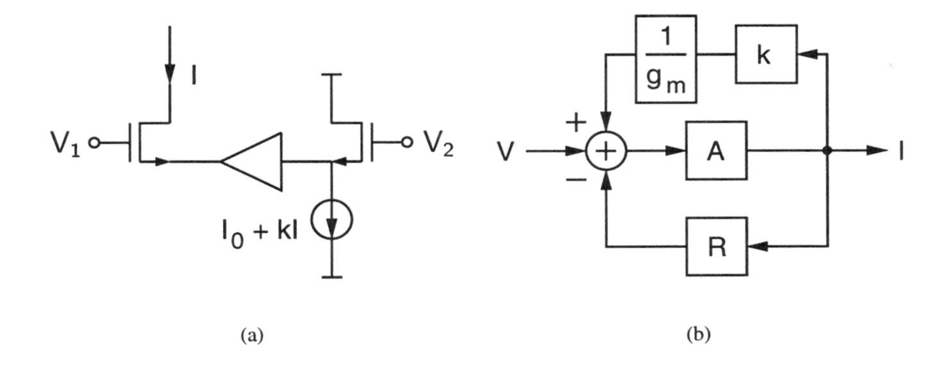

Figure 6.12. Cancelling mobility degradation with positive feedback. (a) Modified transconductance cell. (b) System view.

In contrast to linearization by negative feedback in the class-A case, this approach linearizes by *positive* feedback, increasing both G_m and the linear input range with negligible additional static power consumption.

3.2 A SURVEY OF TRANSCONDUCTOR ARCHITECTURES

To evaluate the merit of this proposed transconductor architecture, we survey several common transconductors and compare their figures-of-merit. For a fair comparison, the same tail current ($I_0 = 1\text{mA}$) biases each transconductor cell, and relevant design parameters are swept in simulation to determine what figures-of-merit are possible with each architecture. Devices with 2-μm channel lengths are used in all simulations. The technology is 0.5-μm CMOS.

The Γ_m parameter can be calculated for each case, using

$$\beta = \frac{G_m}{P_D} \tag{6.33}$$

$$\epsilon = \frac{\overline{i^2_{n,out}}}{4kTBG_m} \tag{6.34}$$

$$V^2_{IP3} = \frac{4}{3}\frac{\alpha_1}{\alpha_3} \tag{6.35}$$

where

$$\alpha_1 = \left.\frac{\partial \Delta I}{\partial \Delta V}\right|_{\Delta V=0} \tag{6.36}$$

$$\alpha_3 = \left.\frac{\partial^3 \Delta I}{\partial \Delta V^3}\right|_{\Delta V=0}. \tag{6.37}$$

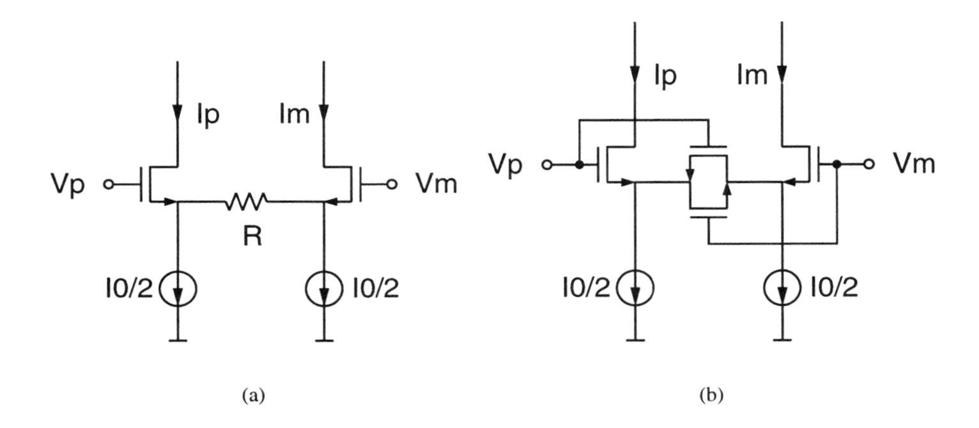

Figure 6.13. Two class-A transconductor architectures. (a) Standard differential pair with resistive degeneration. (b) MOSFET-degenerated differential pair.

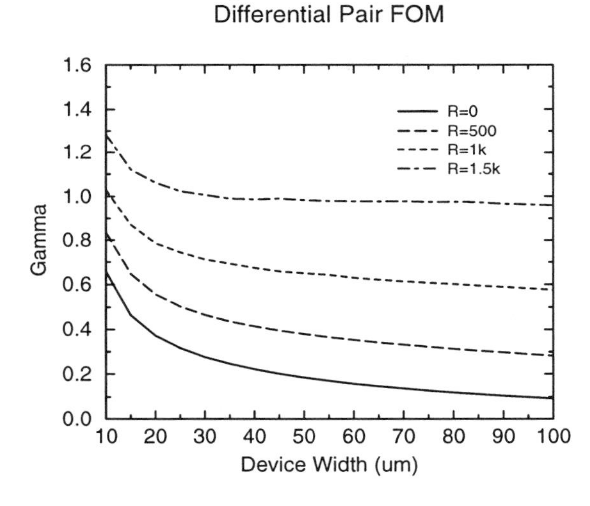

Figure 6.14. Figure of merit for a simple differential pair with source degeneration.

The first of these is the simple differential pair, shown in Figure 6.13(a) To improve Γ_m for a differential pair, one should either use resistive degeneration or operate with high current densities. The maximum current density will be set by the available voltage headroom. For example, with a width of 10μm and a tail current of 1mA, the device in this simulation has a V_{gs} of about 1.8V. Even with such a high V_{gs}, the differential pair has only a modest figure of merit, Γ_m, as shown in Figure 6.14. This plot illustrates the difficulty in obtaining a

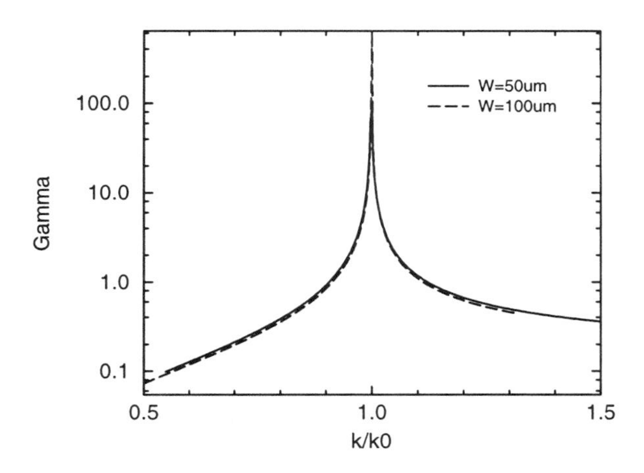

Figure 6.15. Figure of merit for the MOSFET-degenerated differential pair.

Γ_m of greater than unity, even with substantial degeneration. Also, note that β decreases as Γ_m improves, so that the filter impedance increases for a given power dissipation. This increase is undesirable, because larger signal voltages are implied for a higher impedance level.

Another approach, first outlined in [101], is the MOSFET-degenerated differential pair of Figure 6.13(b). This circuit uses variable conductances in the form of triode MOSFET devices to degenerate the differential pair. The simulated Γ_m for this architecture is plotted in Figure 6.15. The degenerating devices are k times the width of the input devices, and the figure shows Γ_m versus k for two device widths. It is interesting that this architecture offers, in principle, virtually unbounded Γ_m with the proper selection of k. This observation stems from the fact that the third-order nonlinearity can be completely cancelled and, since V_{IP3} is projected from small amplitudes, it can be made infinite. Of course, in reality, the transconductor will eventually distort, but such large-signal compression behavior is not captured by Γ_m.

The class-AB transconductor detailed in the previous section is shown in Figure 6.16. Its figures-of-merit are plotted in Figure 6.17. In this figure, k is the positive feedback factor. The simulations show values of Γ_m for two different device widths. Qualitatively, this architecture bears many similarities to the MOSFET-degenerated architecture. The third-order non-linearity can be completely cancelled with proper selection of k. Note, however, that in this case, the skirts of Γ_m are much broader, leading to a larger figure-of-merit if k is misadjusted by a given percentage. Furthermore, the power-efficiency, β, of the class-AB transconductor is 2-3 times larger, leading to lower filter

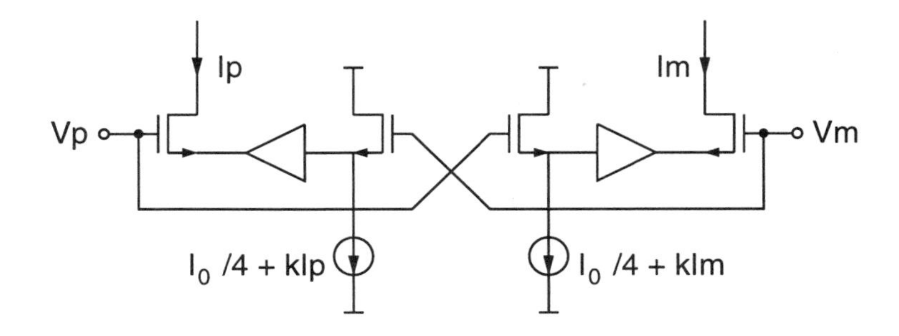

Figure 6.16. The mobility-compensated class-AB transconductor.

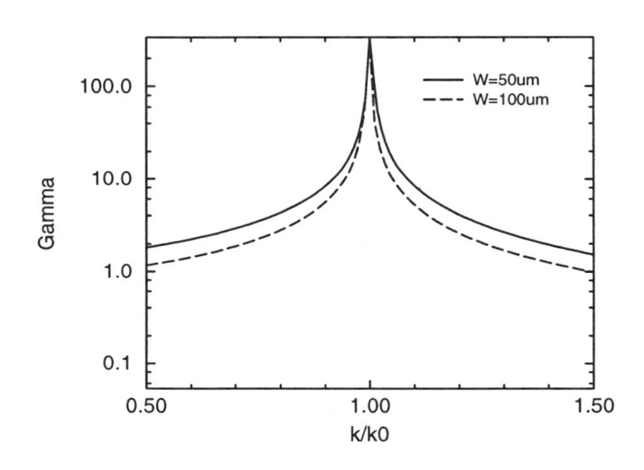

Figure 6.17. Figure of merit for the class-AB transconductor.

impedance levels for a given power consumption. This impedance reduction benefits large-signal handling capability in the filter by reducing the signal voltages for a given source power. Accordingly, we conclude that the class-AB transconductor offers superior performance compared to simple resistive degeneration or MOSFET-degeneration.

3.3 TRANSCONDUCTOR IMPLEMENTATION

A transconductor that applies positive feedback to the task of compensating for vertical field mobility degradation is illustrated in Figure 6.18. The voltage buffer of Figure 6.12 is implemented with M2–M4, which form a linearized, level-shifting buffer, similar to that used in the IFA. Thus, M1 is the

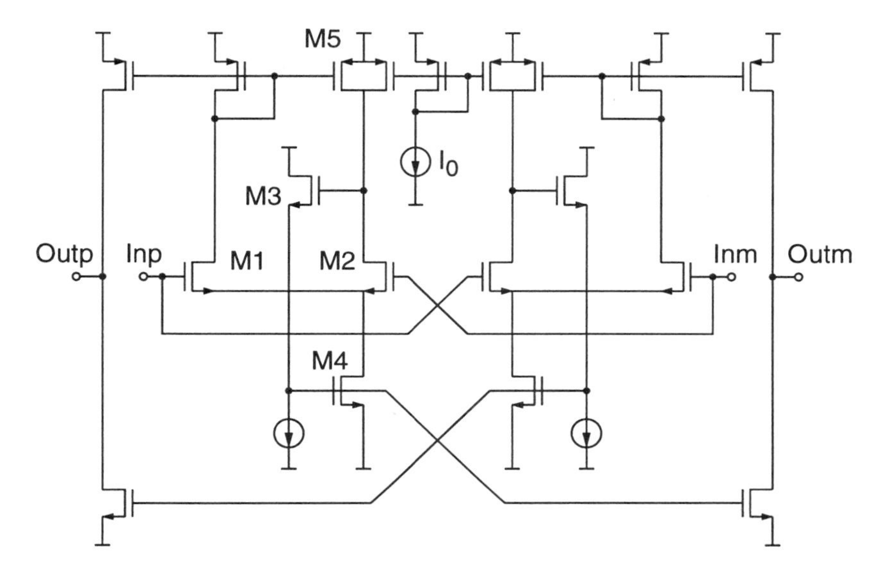

Figure 6.18. A linearized class-AB transconductor.

transconducting NMOS device, and M5 implements the positive feedback path by sampling the output current of M1 and adjusting the bias current to M2.

Figure 6.19 demonstrates the benefits of positive feedback in this architecture. The lower curve shows the nonlinearity in G_m when positive feedback is omitted. The bowing is nearly eliminated with the addition of device M5, and is relatively insensitive to reasonable variation in the length of that device.

Eight of these transconductors are used in the on-chip filter, which has a total power consumption of 9.7mW and a differential terminating impedance of 2kΩ. The filter achieves a peak SFDR of greater than 60dB.

4. SUMMARY

Active filters pose a number of design challenges that have been addressed in this chapter. The goal of wide dynamic range competes directly with the need for selectivity, as has been demonstrated through a dynamic range analysis of the G_m-C architecture. Hence, architectural decisions that relax the required filter order and Q are essential if wide dynamic range is desired on low power consumption.

Much of the dynamic range burden falls naturally on the transconductor that implements the active inductors in the filter. A side effect of the dynamic range analysis is the formulation of a figure of merit for transconductors that elucidates effective approaches to maximizing power-efficiency. Using this figure of merit as a guide, we have developed a new transconductor architecture

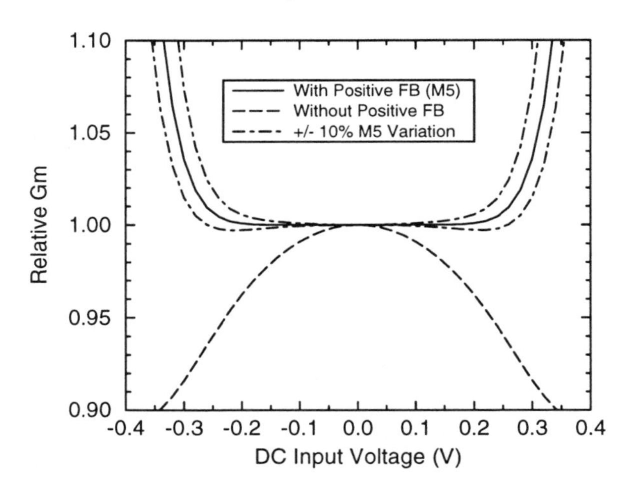

Figure 6.19. Normalized transconductance characteristic, with and without positive feedback.

that employs positive feedback to linearize the transconductance characteristic in a Class-AB transconductor. This development enables the realization of a 10-mW filter that has 60-dB spurious-free dynamic range and a bandwidth of 3MHz. Although the filter has a relatively large dynamic range, it nonetheless limits the dynamic range of the receiver as a whole.

In the next chapter, we present the final implementation of the complete integrated GPS receiver. As will be shown, this receiver achieves a level of performance that compares favorably with existing commercial solutions in superior process technologies.

Chapter 7

AN EXPERIMENTAL CMOS GLOBAL POSITIONING SYSTEM RECEIVER

The previous chapters present design methodologies for some of the critical building blocks in a CMOS radio receiver. To put these ideas into practice, we have implemented a complete GPS receiver that comprises the LNA, mixers, active filters, limiting amplifiers and comparators necessary to render the complete receiver signal path. In addition, the receiver includes an on-chip phase-locked loop (PLL) for synthesis of the first local oscillator. The details of the PLL are presented elsewhere [59].

In this chapter, we explore the implementation and testing of the receiver signal path. A comprehensive description of the signal path and biasing circuit design is presented in Sections 1.–6.. Then, complete experimental results are presented in Section 7.. The chapter concludes with a summary in Section 8..

1. LOW-NOISE AMPLIFIER

The first block in the signal path is the low-noise amplifier. Figure 7.1 shows the complete schematic of the LNA, including biasing circuitry.

The LNA uses a differential architecture to increase its immunity to common-mode interference from substrate or supply perturbations. The input stage, M1–M8, is a cascode amplifier that delivers its output current into a tuned load formed by spiral inductors L3–L4 and the gate capacitances of the second stage, M9–M10. AC coupling between the stages via C1–C2 allows for the bias current in the input stage to be shared with the output stage, thereby reducing power consumption. The second stage drives another pair of tuned loads formed by L5–L6 and the parasitic capacitances at the output nodes.

To bias the LNA, a common-mode feedback (CMFB) circuit drives the voltage at the gates of the cascoding devices, M5–M8, to a desired level. The cascode common mode voltage is measured by R4–R5 and compared to a reference voltage generated by resistive divider R2–R3. These two voltages

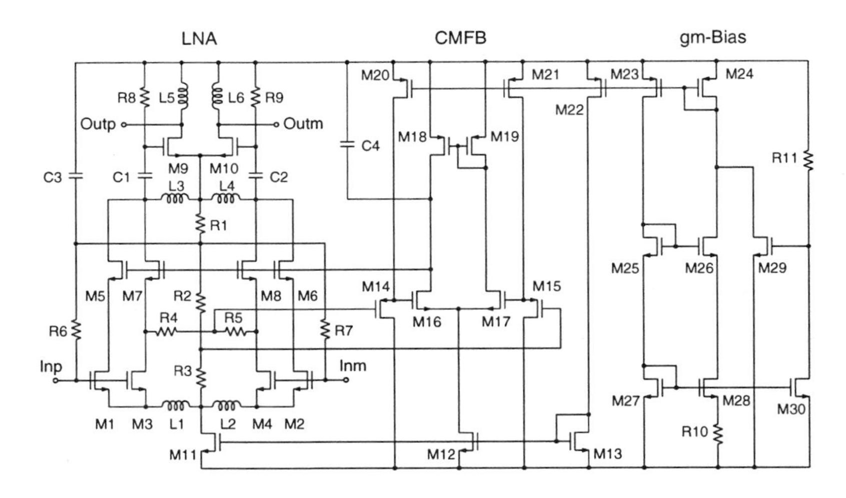

Figure 7.1. The low-noise amplifier.

Table 7.1. LNA Elements

Device	Value	Device	Value
M1–M2	240/0.5μm	M3–M4	20/0.5μm
M5–M6	240/0.5μm	M7–M8	20/0.5μm
M9–M10	240/0.5μm	M11	540/1.2μm
M12–M13	30/1.2μm	M14–M15	120/1.2μm
M16–M17	60/1.2μm	M18–M19	120/1.2μm
M20–M21	30/1.55μm	M22–M24	120/1.55μm
M25–M26	90/1.55μm	M27	30/0.5μm
M28	240/0.5μm	M29	45/1.55μm
M30	30/0.5μm	R1	2.5kΩ
R2–R5	5kΩ	R6–R7	51kΩ
R8–R9	5kΩ	R10	500Ω
R11	20kΩ	C1–C2	2pF
C3–C4	8pF	L1–L2	1.2nH
L3–L4	5.5nH	L5–L6	5.8nH
Total Power	15.75mW	Bias Power	3.5mW

are equalized by feedback action of an operational amplifier, M14–M21. The resistive divider senses the gate-source voltage of the input devices through resistors R6–R7. Thus, the final drain-source voltages of the input devices are proportional to their own gate-source voltages. Capacitor C3 bypasses this resistor network to the supply at radio frequencies.

Because the CMFB technique requires measurement of the source voltages of the cascoding devices, the cascode is split into two parallel branches on each side of the amplifier. This partitioning permits reduced source-drain parasitics in the outer branches because a contact is not required between the input devices and their cascoding counterparts. Thus, the common-mode sensing inner branches use smaller device widths. This reduction in capacitance benefits the noise performance by increasing the high-frequency output impedance of the input devices, in turn reducing the noise contribution of the cascode devices.

Note that all of the voltages in the LNA core are set with respect to the positive supply, so that variations in supply voltage do not affect the quiescent currents in the amplifier. The gates of M9–M10 are set at supply potential so that they can swing *above* the supply when signal is applied, thus making efficient use of the available voltage headroom. The outputs also swing above the supply due to the inductive biasing of the drains of M9–M10.

The bias current for the LNA is provided by a dedicated bias circuit formed by M23–M30 and R10–R11. This type of reference generates a bias current that makes the transconductance, g_{m27}, proportional to a reference conductance, 1/R10. To illustrate the principle of operation, assume that M27 and M28 are square-law devices and that M28 is much wider than M27 so that it operates on a very small overdrive voltage. The positive feedback loop implemented by the current mirror formed by M23-M26 ensures that both M27 and M28 conduct the same current. If we equate the currents in the two branches, we find that

$$\frac{1}{2}\mu_n C_{ox}\frac{W_{27}}{L_{27}}\left(V_{gs27} - V_T\right)^2 = \frac{V_{gs27} - V_{gs28}}{R_{10}}. \tag{7.1}$$

Now, if M28 is a wide device, then it's gate to source voltage is approximately equal to V_T. Then, (7.1) can be simplified:

$$\mu_n C_{ox}\frac{W_{27}}{L_{27}}\left(V_{gs27} - V_T\right) = \frac{2}{R_{10}}. \tag{7.2}$$

The lefthand side of this equation is the transconductance of M27. So, the bias circuit generates the necessary current for the transconductance of M27 to follow 2/R10. This circuit is useful for reducing the variation of the LNA gain and input matching over supply and temperature. In addition, the use of a separate bias cell prevents bias lines from degrading isolation between the PLL and the LNA inputs. Devices M29–M30 and R11 provide a simple startup circuit to eliminate an undesired zero-current state.

2. VOLTAGE-SWITCHING MIXER AND LO DRIVERS

The LNA output drives the inputs of two voltage-switching mixers, driven by quadrature phases of the local oscillator. One of these mixers is shown in Figure 7.2, accompanied by its LO driver.

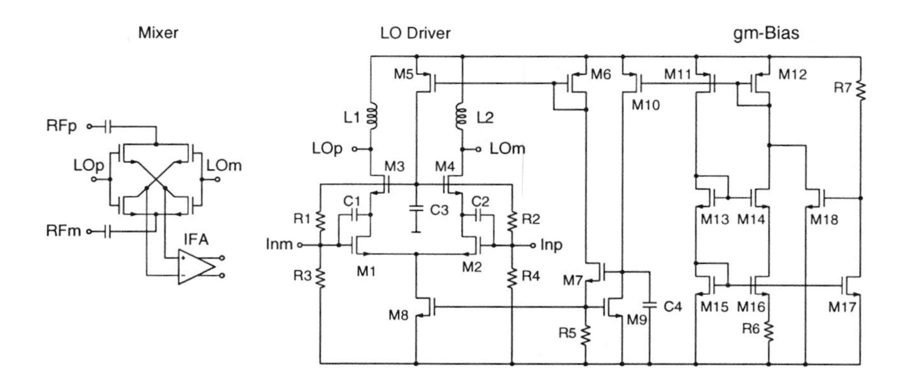

Figure 7.2. Mixer and quadrature LO driver.

Table 7.2. Mixer/LO Driver Elements

Device	Value	Device	Value
M1–M4	80/0.5μm	M5	384/1.2μm
M6	192/1.2μm	M7	48/1.2μm
M8	768/1.2μm	M9	96/1.2μm
M10–M12	120/1.55μm	M13–M14	90/1.55μm
M15	30/0.5μm	M16	240/0.5μm
M17	30/0.5μm	M18	45/1.55μm
R1–R2	20kΩ	R3–R4	40kΩ
R5	20kΩ	R6	500Ω
R7	20kΩ	C1–C2	0.92pF
C3	2pF	C4	6pF
L1–L2	14.2nH		
Total Power	6.83mW	Bias Power	1.67mW

As discussed in Chapter 5, the mixer uses four MOS voltage switches that connect the RF input to the IF output with alternating polarity. The change in polarity is controlled by the local oscillator via the LO driver circuit.

This circuit is a simple cascode amplifier, formed by M1–M4, with inductive loads L1–L2 that resonate the total capacitance at the driver output nodes, including the input capacitance of the four switches in the mixer. Figure 7.2 shows the quadrature version of the LO driver, with C1 and C2 serving to establish the LHP pole and RHP zero that result in a 90° phase shift. The in-phase version of the LO driver omits C1 and C2, and the inputs of the two drivers are directly connected to each other.

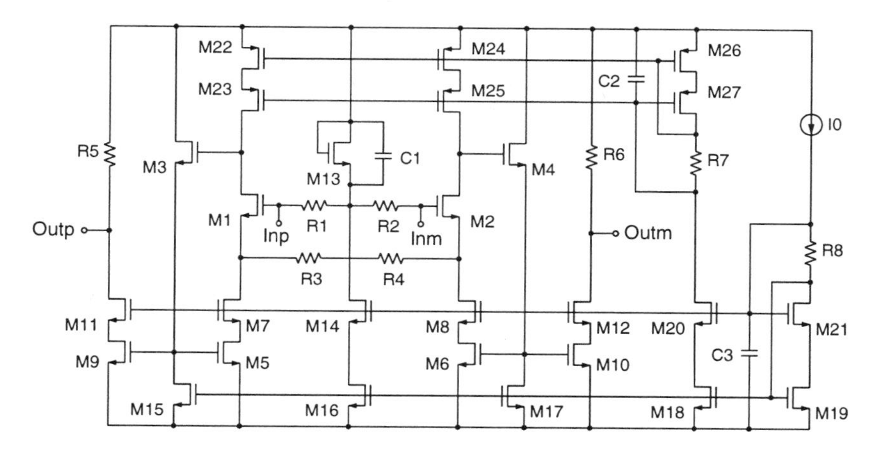

Figure 7.3. Intermediate frequency amplifier.

Because the pole/zero locations depend on g_m, it is important to regulate g_m to be relatively constant so that a constant phase shift is obtained. Thus, the LO driver also uses a self-biased constant-g_m circuit, identical to the one in the LNA. The driver accepts the bias current via a current mirror formed by M7–M9. M9 is relatively wide so that its gate voltage is approximately equal to V_T. Hence, the current through M7 is proportional to V_T. This current provides the bias for the cascode and input devices through a pair of resistor dividers, R1–R4, so that all bias voltages track V_T, thus providing a measure of bias stability. Capacitor C3 bypasses the cascode at radio frequencies.

3. INTERMEDIATE FREQUENCY AMPLIFIER

The output of the mixer is directly connected to the input of the intermediate frequency amplifier (IFA). To minimize signal currents flowing in the mixer switches, the IFA should present a relatively high input impedance. In addition, the output impedance of the IFA should be well-defined for use as a termination resistor for the active filter that follows. Finally, the IFA should make efficient use of its bias current while providing a linear transfer characteristic, as discussed in Chapter 6.

Figure 7.3 shows an amplifier that meets these requirements. Devices M1-M8 form a linearized voltage buffer that, through feedback action, causes a constant current to flow from drain to source in devices M1–M2. Thus, the input voltage experiences a nearly constant level shift and is placed across the input resistors, R3–R4. This remains true for any input amplitude until the peak current flowing in the resistors is equal to the bias current supplied by

Table 7.3. IFA Elements

Device	Value	Device	Value
M1–M6	96/1.2μm	M7–M8	48/1.2μm
M9–M10	192/1.2μm	M11–M12	96/1.2μm
M13	60/0.5μm	M14	12/1.2μm
M15	48/1.2μm	M16	12/1.2μm
M17–M21	48/1.2μm	M22–M27	192/1.2μm
R1–R2	15kΩ	R3–R4	250Ω
R5–R6	1kΩ	R7	1.5kΩ
R8	2kΩ	C1	4pF
C2–C3	1pF	I0	243μA
Total Power	6.18mW	Bias Power	1.2mW

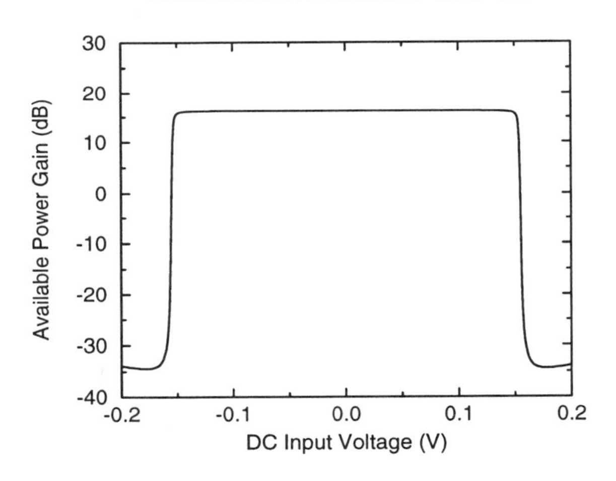

Figure 7.4. Simulated gain characteristic of the IFA.

M22–M25. Beyond this point, the feedback breaks down, and the amplifier saturates.

Because the linear signal current through R3–R4 must flow in devices M5–M6, it is a simple matter to mirror this current to the output loads with devices M9–M10. The feedback results in a very linear gain characteristic with a sharp saturation behavior, as illustrated in Figure 7.4, which shows the simulated available power gain of the amplifier. Because there is only one pair of high-impedance nodes in the current feedback loop, the amplifier does not require frequency compensation, and its step response exhibits no overshoot.

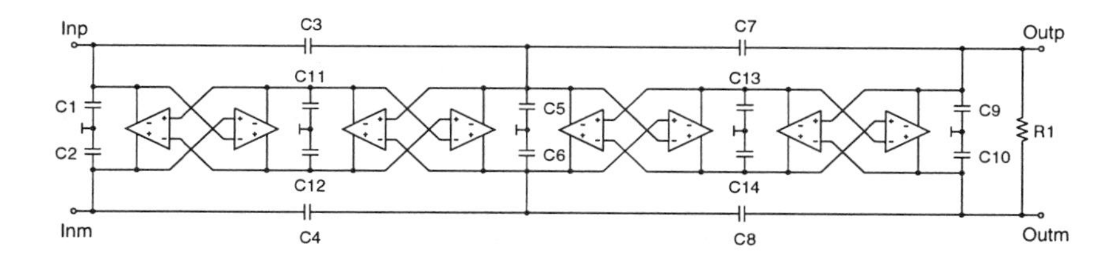

Figure 7.5. Active G_m–C filter. The missing input termination resistor is supplied by the output resistance of the preceding IFA stage.

Table 7.4. Filter Capacitors

Capacitors	600fF Units (ea.)	Value (ea.)
C1–C2	84	50.4pF
C3–C4	2	1.2pF
C5–C6	173	103.8pF
C7–C8	5	3.0pF
C9–C10	81	48.6pF
C11–C12	144	86.4pF
C13–C14	139	83.4pF
Total	1256	753.6pF

Note that, because the mixer is dc coupled to the IFA input, the bias circuit that sets the dc voltage for devices M1–M2 also serves to set the dc potential for the mixer devices in Figure 7.2. Device M13 in Figure 7.3 conducts a small current so that the mixer devices are biased approximately one threshold voltage below the positive supply. The choice of bias voltage optimizes the switching point of the mixer devices, B_{LO}, as described in Chapter 5. Capacitor C1 bypasses the bias device at high frequencies.

4. ACTIVE FILTER

The output of the IFA directly drives the input of the active filter, whose schematic is shown in Figure 7.5. In this figure, the required common-mode feedback circuitry has been omitted for clarity. Each pair of differential nodes in the filter, with the exception of the two input nodes, requires common-mode regulation, as will be discussed shortly.

The filter is based on an elliptical L-C ladder prototype with a cutoff frequency of 3.5MHz. The inductive elements of the filter are implemented with

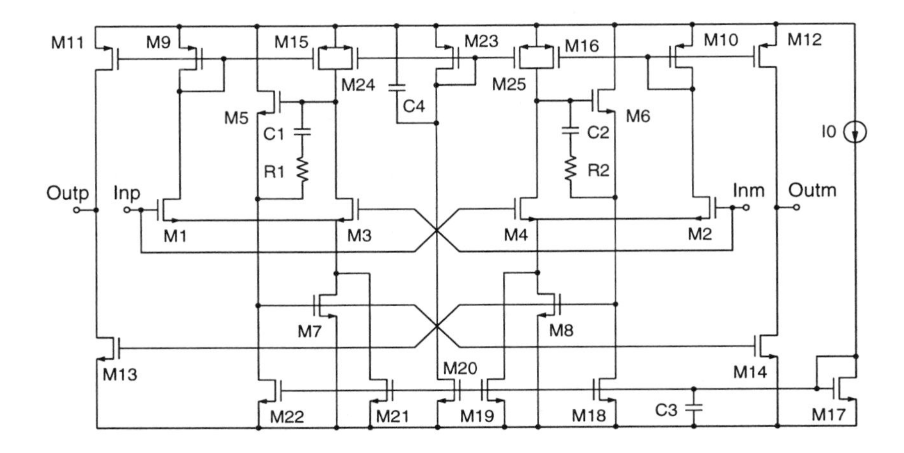

Figure 7.6. Gyrator transconductor.

Table 7.5. Transconductor Elements

Device	Value	Device	Value
M1–M4	5/1.55μm	M5–M6	25/1.2μm
M7–M8	50/1.2μm	M9–M12	25/1.2μm
M13–M14	50/1.2μm	M15–M16	5/1.2μm
M17	18/1.2μm	M18–M19	15/1.2μm
M20	18/1.2μm	M21–M22	15/1.2μm
M23–M25	25/1.2μm	R1–R2	1kΩ
C1–C2	0.8pF	C3–C4	5pF
Total Power	1.07mW	Bias Power	0.18mW

dual gyrators and capacitors C11–C14. All of the capacitors in the filter, with the exception of bridging capacitors C3–C4 and C7–C8, are tied to ground so that they can serve as common-mode feedback compensation.

The transconductor used for the gyrators is shown in Figure 7.6. Devices M1–M2 are the square-law transconductors that are driven by linearized level-shifting buffers M3–M8. The output currents from M1–M2 are sampled and fed-back to the buffers via M15–M16. The feedback factor is adjusted to approximately cancel the mobility degradation nonlinearity in M1–M2, thus linearizing the transconductor.

The linearized output currents are mirrored to the transconductor outputs via M11-M12. In addition, the signal currents in M7–M8 are mirrored to pull-down devices M13–M14. This push-pull action boosts the transconductance by a factor of two without additional power consumption. Devices M19 and M21

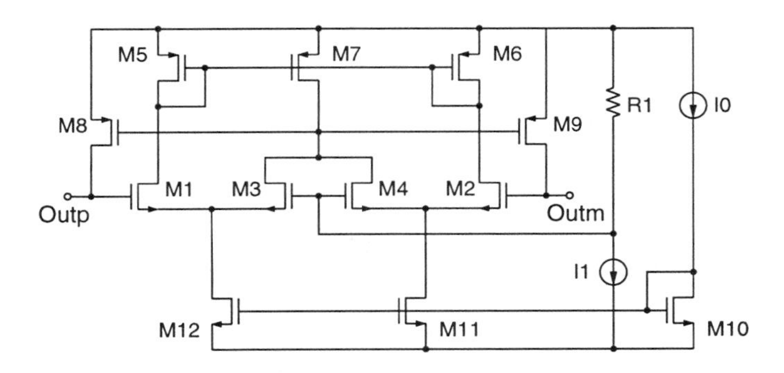

Figure 7.7. Common-mode feedback circuit for the transconductor.

Table 7.6. Transconductor CMFB Elements

Device	Value	Device	Value
M1–M4	5/0.5μm	M5–M6	10/1.2μm
M7	20/1.2μm	M8–M9	80/1.2μm
M10	10/2.6μm	M11–M12	20/2.6μm
R1	33.75kΩ	I1	27μA
Total Power	81μW	Bias Power	16μW

provide static bias current to reduce the standing current in M7 and M8, thereby adjusting the static imbalance in the currents in the output drivers, M11–M12 and M13–M14.

This differential transconductor requires common-mode feedback. A simple CMFB circuit is shown in Figure 7.7. The common-mode voltage at the transconductor outputs is measured with two differential pairs, M1–M4, whose output currents are summed. Differential outputs produce cancelling error currents in the two transconductors; thus, only common-mode disturbances produce a net error. This error is fed back to the gates of PMOS devices M8–M9, adjusting the total pull-up current of the transconductor to regulate the common-mode voltage. The reference voltage for this circuit is derived from a bandgap-reference current, I1, which produces a nearly constant voltage across resistor R1. To frequency-compensate the CMFB circuit, the filter capacitors are referenced to ground so that a dominant pole is established at the CMFB output nodes.

To generate the appropriate bias current for the filter, we can borrow the concept of using positive feedback to slave g_m to a reference resistor. In this case, however, a replica transconductor is used in a feedback loop so that its

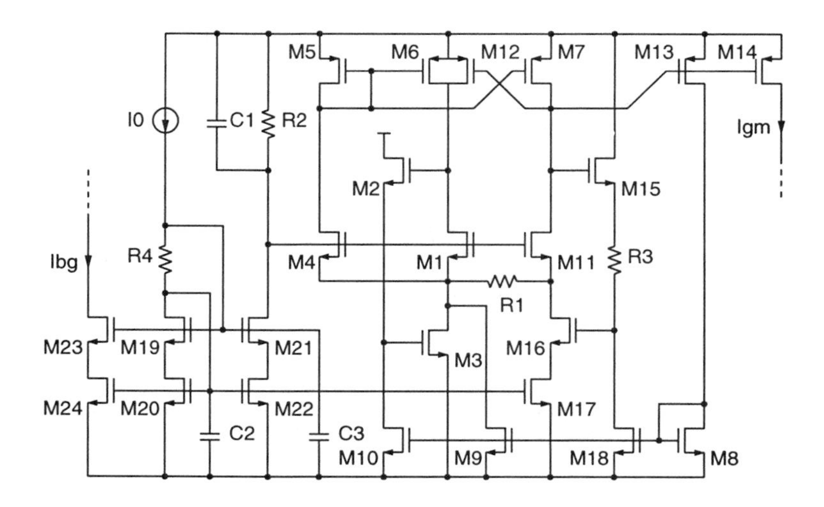

Figure 7.8. Replica biasing of the filter transconductor.

Table 7.7. Replica Bias Elements

Device	Value	Device	Value
M1	5/1.55μm	M2	25/1.2μm
M3	50/1.2μm	M4	5/1.55μm
M5	25/1.2μm	M6	5/1.2μm
M7	25/1.2μm	M8	18/1.2μm
M9–M10	15/1.2μm	M11	40/1.55μm
M12–M14	25/1.2μm	M15–M16	50/1.2μm
M17	12/1.2μm	M18	15/1.2μm
M19–M24	12/1.2μm	R1	5.8kΩ
R2	33.75kΩ	R3	40kΩ
R4	20kΩ	C1	40pF
C2–C3	20pF	I0	27μA
Total Power	911μW	Bias Power	—

total transconductance, G_m, is slaved to a reference resistor, R. A circuit that achieves this purpose is shown in Figure 7.8.

This circuit accepts a single reference current, I0, from the bandgap reference, and produces two output currents, Ibg and Igm. The first of these is just a scaled version of the bandgap current for use in the CMFB circuitry in the filter. The second has a more complex origin that requires a detailed explanation.

Devices M1–M10 are a half-circuit replica of a single transconductance stage in the filter. This replica is in a positive feedback loop with M11 and R1, and the

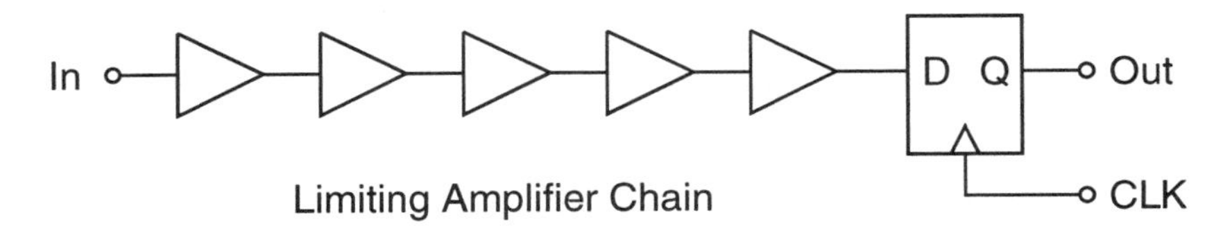

Figure 7.9. Five-stage limiting amplifier and output comparator.

loop is closed with devices M12–M13, which set the bias current for the whole circuit. Due to the positive feedback, this circuit finds a stable operating point when the total bias current reaches a level that causes the transconductance of the replica to be proportional to the reference resistor, R1. Note that resistor R2 sets the gate potential of M1 and M4 to the nominal common-mode voltage of the filter, so that body effect is accounted for. The resulting bias current is supplied to the filter transconductors via current mirror taps, as shown with M14 in the figure. Because the termination resistors in the filter are made of the same material as the reference resistor, the filter shape is well regulated by this circuit. There will, however, be variation in the resulting filter cutoff frequency as the RC product changes with process. Because the RC product may change by $\pm 15\%$, the filter must be designed to have a slightly larger nominal bandwidth than absolutely necessary. In this design, the filter actually has a nominal bandwidth of 3.5MHz. With process variations, the bandwidth may drop as low as 3.0MHz or rise as high as 4.0MHz. The minimum sampling frequency is thus 8MHz to ensure that no unwanted aliasing occurs.

The replica bias circuit has an undesirable stable state in which the drain of M7 sits at the positive supply potential, thereby shutting off the output current, Igm. Startup circuitry comprising M15–M18 detects and eliminates this unwanted state. Device M16 is a switch that connects a fraction of the reference bandgap current via M17 to the source of M11 during startup. In the zero-current state, the gate of M15 is at the positive supply potential, and M18 is off. Thus, M15 pulls up on the gate of M16, turning it on and passing current from M17 to the source of M11. As M17 pulls down on M11, the circuit begins to start up. Once a stable state has been reached, M18 pulls down on the gate of M16 through R3, shutting off M16 and depriving M11 of the bandgap current. This action effectively removes the startup circuit from operation, leaving the replica circuit unperturbed after the startup period is complete.

5. LIMITING AMPLIFIER AND COMPARATOR

The final two stages in the receiver signal path are the limiting amplifier and output comparator, shown in Figure 7.9.

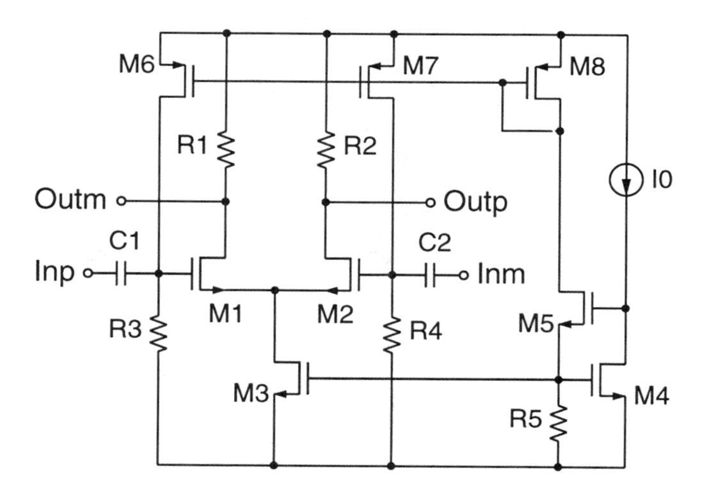

Figure 7.10. A single stage of the limiting amplifier.

Table 7.8. Limiting Amplifier Elements

Device	Value	Device	Value
M1–M2	24/1μm	M3	90/2μm
M4–M5	5/2μm	M6–M7	8/2μm
M8	5/2μm	R1–R2	30kΩ
R3–R4	90kΩ	R5	100kΩ
C1–C2	3pF	I0	5μA
Total Power	305μW	Bias Power	9.2μW

The limiting amplifier is a five-stage amplifier that uses simple differential pairs. The stages are ac coupled to one another to prevent the propagation of dc offsets through the chain, with the exception of the very first stage, which is dc coupled to the filter output. With a 2-MHz IF frequency, the lower pole of the ac coupling should be somewhat below 1MHz to prevent distortion of the C/A code main lobe. The limiting amplifier nominally provides 96dB of voltage gain and 78dB of available power gain, which is more than sufficient to amplify system thermal noise up to a detectable level for the comparator that follows. Any gain in excess of this value would be wasted. A schematic of one of the limiting amplifier stages is shown in Figure 7.10.

The comparator is a standard Yukawa latch [102] that accepts a single clock supplied from off chip and that is driven directly by the output stage of the limiting amplifier. The comparator is clocked at about 16MHz for

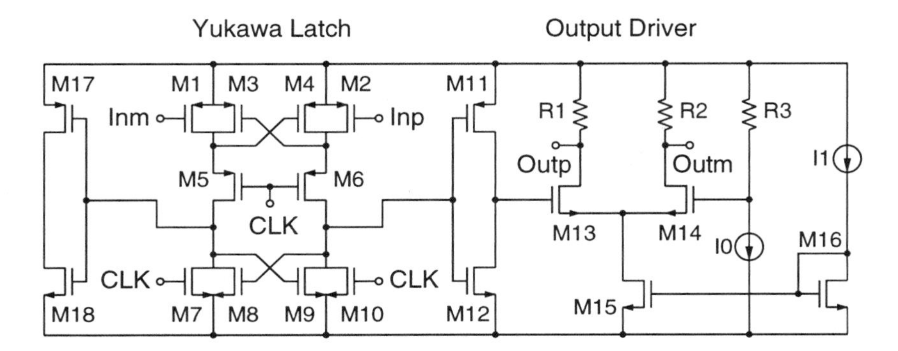

Figure 7.11. The output latch and output driver.

Table 7.9. Latch / Output Driver Elements

Device	Value	Device	Value
M1–M2	48/1.55μm	M3–M4	40/1.55μm
M5–M6	100/1.55μm	M7	20/1.55μm
M8–M9	32/1.55μm	M10	20/1.55μm
M11	10/1.55μm	M12	5/1.55μm
M13–M14	150/1.55μm	M15	120/1.55μm
M16	30/1.55μm	M17	10/1.55μm
M18	5/1.55μm	R1–R2	400Ω
R3	38.5kΩ	I1	243μA
Total Power	3.05mW	Bias Power	0.6mW

an oversampling factor of about two. The output of the comparator drives a differential output driver with on-chip 400Ω loads. The small differential swings at the driver outputs mitigate possible interaction between the output driver and the sensitive LNA input due to substrate coupling. The schematic of the comparator and output driver is shown in Figure 7.11

Note that although the vast majority of system gain occurs in these two blocks, they occupy less than one eighth of the total die area of the chip. No signs of instability were observed, despite the large gain.

6. BIASING DETAILS

With the exception of the LNA, filter and LO drivers, all of the receiver blocks derive their bias from one of two on-chip bandgap references. A simplified circuit diagram of the bandgap reference is shown in Figure 7.12.

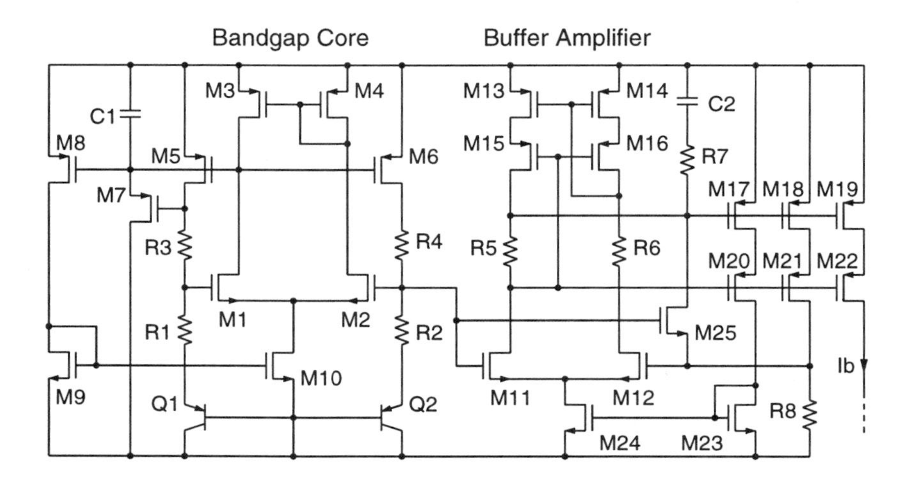

Figure 7.12. Bandgap reference circuit.

Table 7.10. Bandgap Reference Elements

Device	Value	Device	Value
M1–M6	80/1.2μm	M7	60/0.5μm
M8	80/1.2μm	M9	40/1.2μm
M10	80/1.2μm	M11–M12	40/1.2μm
M13–M18	60/1.2μm	M19	180/1.2μm
M20–M21	60/1.2μm	M22	180/1.2μm
M23	20/1.2μm	M24	40/1.2μm
R1	6.25kΩ	R2	6.92kΩ
R3–R4	4.5kΩ	R5–R6	6kΩ
R7	150Ω	R8	15kΩ
C1–C2	16pF	Ib	243μA
Total Power	1.87mW	Bias Power	—

Two substrate PNP devices, Q1–Q2, arrayed with 8:1 emitter area ratios, provide a proportional-to-average-temperature (PTAT) voltage reference that causes a PTAT current to flow in resistors R1 and R2. An active current mirror, using devices M1–M6, ensures that the two current branches conduct identical PTAT currents. This structure is preferable to a passive current mirror, which suffers from inherent current offsets, thus degrading the stability of the reference. In this circuit, as long as M3–M6 possess the same current densities, the amplifier will have no systematic offset. This condition is ensured by devices M8–M10, which bias the operational amplifier with the same PTAT current produced by the bandgap. All systematic offsets are thereby eliminated. Because this is a self-biased technique, the threat of a zero-current state must be removed by appropriate startup circuitry. Device M7 and resistor R3 serve this purpose, and resistor R4 balances the amplifier so that the drain voltages of M5–M6 are nominally equal.

The voltage appearing at the gate of M2 is the sum of one V_{be}, which is comple-mentary-to-average-temperature (CTAT), and a PTAT voltage across R2. By proper ratioing of R1 and R2, the resulting voltage is constant with temperature. The static supply rejection of this circuit is comparable to the temperature variation of the bandgap voltage due to the balance maintained by the active current mirror.

A buffer amplifier accepts the bandgap voltage as an input and buffers it across a resistor, R8, thus producing a current that is proportional to the bandgap voltage that can be supplied to various receiver circuit blocks. Just as in the bandgap core itself, attention to systematic offsets is crucial for providing a stable current.

Again, to balance out any systematic offsets, the buffer amplifier uses its own output current as its tail current. Thus, M13–M19 can be designed to conduct identical current densities, equalizing the drain voltages of M15–M16. Resistor R5 provides a level shift to generate the cascode gate voltages for M20–M22, and R6 serves to equalize the drain voltages of M11 and M12. Due to the use of self-biasing in this amplifier, a startup circuit is again required, and a simple solution is found in M25, which initializes current flow through R8 and pulls down on the gate of M17 to start the amplifier. Elements C1–C2 and R7 provide loop stability compensation for the bandgap core and buffer amplifier.

This bandgap circuit provides 27-μA and 243-μA bias current taps for distribution throughout the receiver.

7. EXPERIMENTAL RESULTS

The GPS receiver has been implemented in a 0.5-μm CMOS process and a die micrograph is shown in Figure 7.13. The layout occupies 11.2mm^2 and uses 16 spiral inductors in the RF and PLL sections. These spirals use patterned ground shields for improved quality factor and reduced crosstalk between spirals [52]. A comparison between simulations and measurements of the spiral inductors is shown in Table 7.11, indicating excellent agreement. The discrepancies for the second and third inductors in the table arise mostly from inconvenient connecting stubs on the test structures whose parasitic contributions are not de-embedded. The simulation model used in this work is described in [103].

The entire signal path of the chip is differential, and careful attention is paid to symmetry throughout the layout. The I and Q channels are also symmetrically placed about the horizontal centerline of the chip. To reduce interaction between

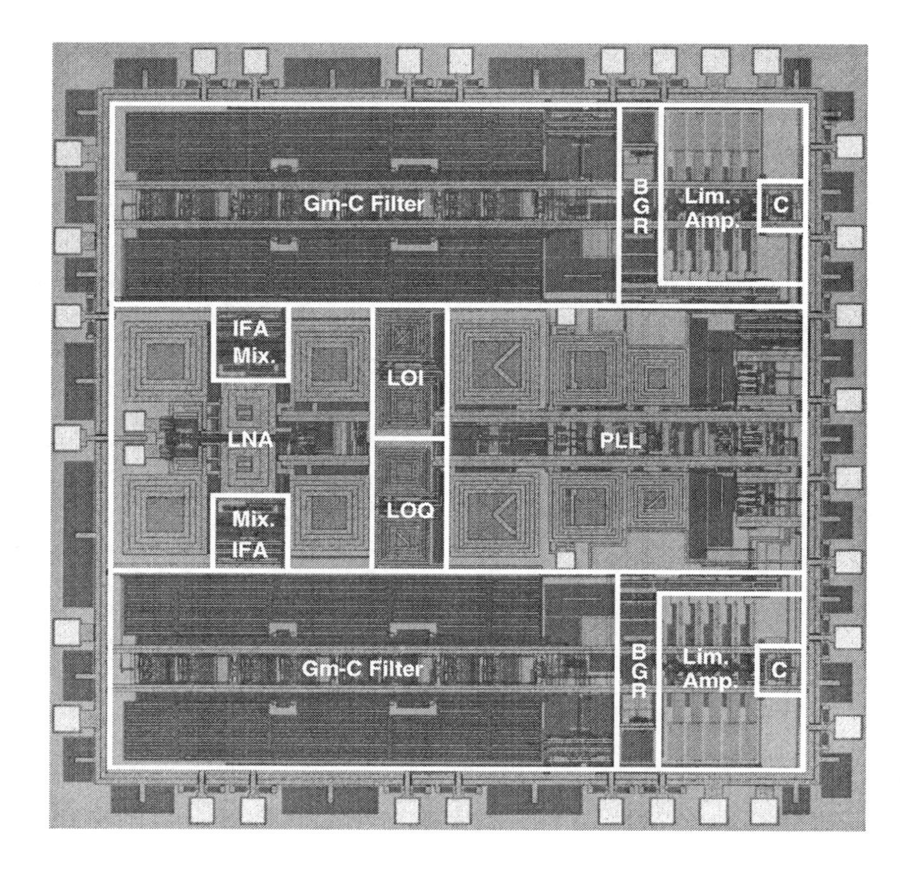

Figure 7.13. Die micrograph of the GPS receiver.

Table 7.11. Spiral Inductors

Simulated Inductance	Q	Measured Inductance	Q
1.2 nH	6.7	1.4 nH	6.8
5.6 nH	7.6	4.8 nH	6.6
5.9 nH	7.4	5.6 nH	6.6
6.9 nH	7.0	6.6 nH	6.8
10.0 nH	4.7	9.5 nH	4.5
10.0 nH	6.3	10.3 nH	6.0
14.3 nH	5.1	14.5 nH	5.2

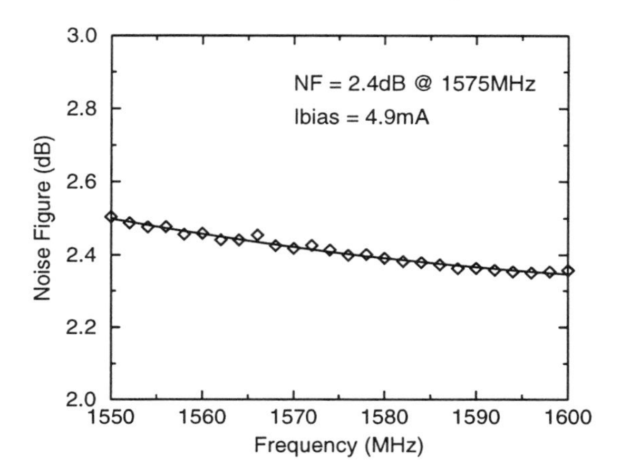

Figure 7.14. Measured LNA noise figure.

the LNA and PLL circuitry, separate supplies are run from the outer supply ring, where extensive on-chip capacitive bypassing is used (about 1.2nF, in all).

The LNA is also laid out as a separate test structure so that its noise figure can be measured independently. The result is shown in Figure 7.14. The LNA has a noise figure of 2.4dB at 1.575GHz with 4.9mA of bias current in the amplifier core, suggesting that an equivalent single-ended amplifier would consume 2.45mA of bias current. The input return loss was better than 20dB for this noise figure measurement. Based on simulations of the LNA with models that include the induced gate noise and that have been verified against S-parameter measurements, this noise figure corresponds to a γ of approximately 1.2. Note that this value of γ, while greater than the long-channel value of 2/3, is actually smaller than values reported in [2]. One possible explanation is that lightly-doped drain (LDD) structures in modern MOSFET devices help to reduce the electron temperature near the drain end of the channel, thereby controlling the increase in γ [87].

Several test points are located on the die for measuring intermediate points along the signal path. In particular, an output buffer amplifier permits signal path measurements after the filter and before the limiting amplifier. Most of the signal path data are based on measurements at this test point.

Figure 7.15 shows the signal path frequency response as measured before the limiting amplifier. The simulated and measured responses agree very well. The filter exhibits about 77dB of stopband rejection and less than 1dB of passband peaking. Mismatch in I and Q amplitudes is observed, some of which is attributed to board components. Unfortunately, a direct measurement of the

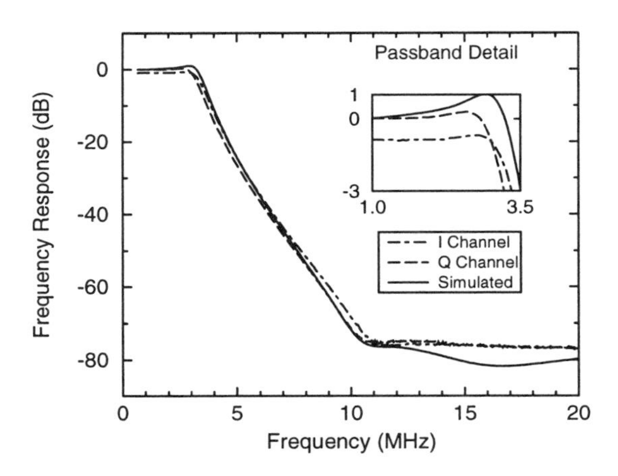

Figure 7.15. Measured signal path frequency response.

on-chip I and Q matching is not possible, but the success or failure of image rejection can be readily ascertained by measuring the post-detection SNR.

A spot noise figure of 2.8dB is also measured at the output of the filter. This number is in good agreement with the predicted value of 2.9dB from design simulations of the individual receiver blocks. The use of a 1-bit quantizer causes an SNR degradation that can be accounted for by increasing the effective noise figure to 4.5dB.

To determine the linearity of the system, two tests are performed: a two-tone IM3 test, and a 1-dB blocking desensitization test. These are shown in Figures 7.16 and 7.17.

For the IM3 test, two in-band test tones are applied to the system at 1.57562GHz and 1.57542GHz. Note that the classical behavior of the IM3 products breaks down above an available source power of –51dBm. The subsequent rise in distortion may be attributed to the rapid increase in G_m observable in the filter transconductor characteristic of Figure 6.19 when the input amplitude exceeds a certain value. Because the received signal power in the GPS system is very low, an extrapolation from low source powers is most relevant, yielding a –25-dBm input referred IP3. This number is set almost entirely by distortion in the active filter and could thus be improved at the expense of increased filter power consumption.

A more relevant performance measure for this system is the 1-dB blocking point. In Figure 7.17, a single out-of-band blocker is applied to the system, and its power is increased until a 1-dB reduction in the in-band SNR is observed. The band of frequencies that presents the greatest blocking threat is

Signal Path 3rd Order Intermodulation

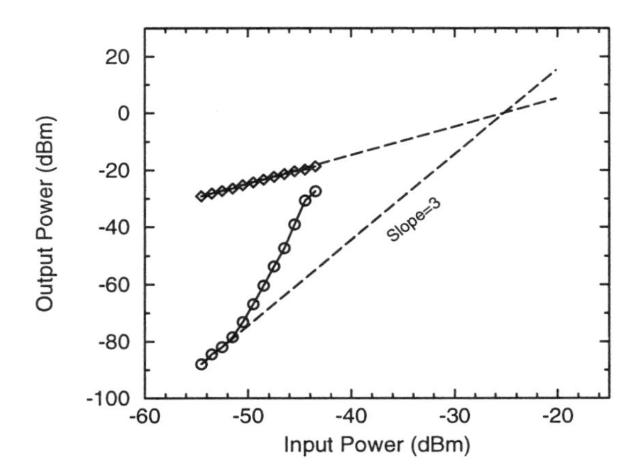

Figure 7.16. Results of a two-tone IM3 test.

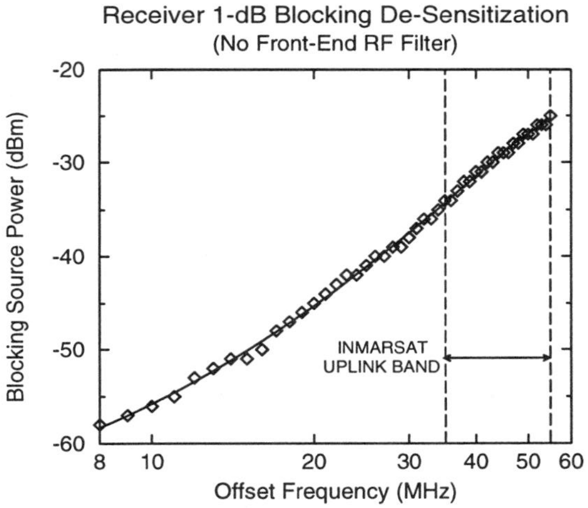

Receiver 1-dB Blocking De-Sensitization
(No Front-End RF Filter)

Blocking Source Power (dBm)
-20
-30
-40
-50
-60
INMARSAT
UPLINK BAND
8 10 20 30 40 50 60
Offset Frequency (MHz)

Figure 7.17. Measured 1-dB blocking desensitization point.

the INMARSAT uplink band, positioned at an offset of 35-MHz to 55-MHz from the GPS center frequency of 1.575GHz. At the lower edge of this band, the receiver has a 1-dB blocking point of –35-dBm available source power. Note that no external RF filtering is used in making this measurement. With a reasonable filter, this number would improve by 15–20dB in the final system. The measured blocking performance is consistent with a PLL phase noise of better than –135dBc/Hz at 35-MHz offset.

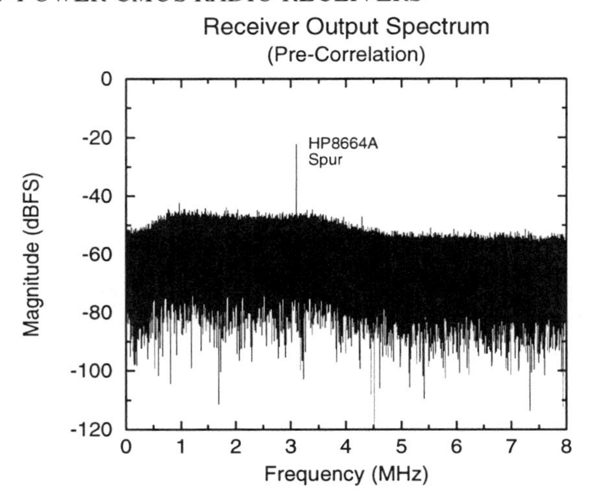

Figure 7.18. FFT of the I channel output bit sequence.

With the signal path and PLL verified separately, the full receiver is finally tested with a simulated GPS signal applied to the input at –130-dBm available source power. The 1-bit digital output stream is captured for the I and Q channels and digitally downconverted using a *noncoherent* back end to reduce the computational complexity. The use of a noncoherent back end causes an additional SNR degradation that elevates the effective noise figure to about 6.7dB.

A fast Fourier transform (FFT) of the I channel is shown in Figure 7.18. Despite the de-correlating effects of the limiter, the filter characteristic is still visible as an increase in the noise floor from 1–3MHz. The spectrum is free of spurs, with the exception of a single spur attributable to one of the bench-top references used for the PLL.

The downconverted bit sequences are correlated with a reference copy of the GPS spreading code and the resulting cross-correlation as a function of code phase is plotted in Figure 7.19. By comparing the magnitude of the correlation peak with the variance of the off-peak cross-correlation, one concludes that the output SNR is about 17dB. Recalling that a received SNR of –19dB is expected, with a processing gain of 43dB and an effective noise figure of 6.7dB, we expect

$$SNR = -19\text{dB} + 43\text{dB} - 6.7\text{dB} = 17.3dB \tag{7.3}$$

which agrees very well with the measured result of 17dB. The conclusion is that the I/Q matching in the system is sufficient for effective cancellation of the image noise. Table 7.13 summarizes the receiver performance.

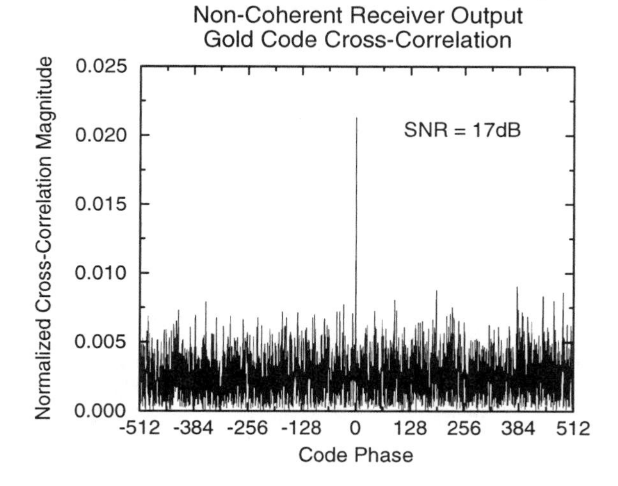

Figure 7.19. Cross-correlation at the receiver output.

Table 7.12. Comparison with Commercial GPS Receivers.

Specification	This Work	Sony (JSSC Apr'97)	GEC Plessy GP2010	SiRF GRF-1
LNA NF	2.4dB (int)	2dB (ext)	≈ 2dB (ext)	(ext)
Chip NF	2.9dB	6.1dB	≈ 10dB	—
IIP3	-25dBm	-14.5dBm	—	—
Avail. Gain	103dB	107dB	106dB	—
ADC	1-bit	1-bit	2-bit	2-bit
Power	115mW @ 2.5V	81mW @ 3V	200mW @ 3V	500mW @ 5V
Technology	0.5μm CMOS	15GHz Bipolar	Bipolar	—
Missing	—	2 Filters LNA, PLL LF	2 Filters, LNA, PLL LF	Filter, LNA

In addition, Table 7.12 presents a comparison between the experimental results of the present work and several commercial GPS receivers that are available today. This comparison illustrates that the CMOS GPS receiver achieves similar or better performance in a less expensive technology with more complete system integration onto a single chip.

8. SUMMARY

In summary, this chapter describes the implementation of a complete CMOS GPS receiver that includes all necessary active blocks in the RF and analog signal path, plus a PLL for LO synthesis. The details of the PLL are presented

Table 7.13. Measured GPS receiver performance.

Signal Path Performance	
LNA Noise Figure	2.4dB
LNA S11	–20dB
Coherent Receiver NF	2.8dB
IIP3 (Filter-limited)	–25dBm @ –51dBm P_s
Peak SFDR	56dB
Filter Cutoff Frequency	3.5MHz
Filter Passband Peaking	$\leq$ 1dB
Filter Stopband Attenuation	$\geq$ 52dB @ 8MHz
	$\geq$ 68dB @ 10MHz
Pre-Filter Avail. Power Gain	28dB
Pre-Filter Voltage Gain	41dB
Total Avail. Power Gain	$\approx$ 103dB
Total Voltage Gain	$\approx$ 131dB
Non-Coherent Output SNR	17dB
LO Leakage @ LNA Input	$<$ –72dBm
Power Dissipation	
Signal Path	79mW
PLL / VCO	36mW
Supply Voltage	2.5V
Implementation	
Die Area	11.2mm^2
Technology	0.5-μm CMOS

elsewhere [59]. The signal path successfully applies the low-IF architecture by exploiting details of the GPS signal structure to permit a reduction in the I/Q matching requirements and a relaxation of the channel filtering problem.

The final system consumes 115mW from a 2.5-V power supply and occupies 11.2mm^2 of die area in a 0.5-μm CMOS process. It is capable of detecting a –130-dBm GPS signal with a noncoherent back-end SNR of 17dB.

Chapter 8

CONCLUSIONS

This book has examined in detail a number of issues related to integrated radio receivers, particularly in the context of CMOS technologies. The techniques presented in earlier chapters have enabled the implementation of a low-power, high-performance CMOS GPS receiver in a 0.5μm technology. This receiver consumes less power and yields comparable or better performance with a higher level of integration than most commercial receivers available today. This demonstrates that CMOS technologies are viable alternatives to more expensive silicon bipolar and GaAs MESFET technologies commonly employed for integrated receiver applications.

To conclude, we now briefly summarize the key contributions presented in the previous chapters.

1. SUMMARY

The vast majority of integrated GPS receivers use a standard superheterodyne architecture with a number of off-chip components, particularly passive IF filters. In Chapter 3, it was demonstrated how the detailed nature of the GPS signal spectrum presents an opportunity for a low-IF receiver architecture that offers the benefit of complete integration of the receiver signal path. Because the I/Q matching requirements are relaxed in this architecture, adequate image rejection is readily obtained without trimming or calibration.

With the goal of minimum power consumption in mind, subsequent chapters set out in detail how to optimize the design of the signal path elements. Beginning with the low-noise amplifier, Chapter 4 demonstrates the theoretical limits to low-noise operation in CMOS by including an oft-neglected noise source: induced gate noise. By optimizing the device geometry under the constraint of a good input match and constant power consumption, the conditions for minimum noise figure were derived. This "power-constrained" optimization demonstrates that excellent noise performance can be achieved in CMOS with small numbers of milliwatts while delivering a good input impedance match to the off-chip 50Ω world. This theoretical development enables a 2.4dB noise figure for a differential LNA with only 4.9mA of bias current.

The design of the second signal path element, the mixer, is the subject of the next chapter. The double-balanced CMOS voltage mixer consumes exceptionally little power with no bias current required in the mixer core. Although conversion gain is a concern in this architecture, a careful analysis demonstrates that higher conversion gains can be obtained by reactively terminating the IF port of the mixer. A noise figure analysis further demonstrated that SSB noise figures on the order of 6dB are easy to obtain. Finally, the linearity of the mixer is primarily limited by the magnitude of the LO drive. Thus, the CMOS voltage mixer achieves wide dynamic range and excellent noise figure with no static power consumption.

The next critical signal path block is the on-chip active filter. Because active filters are automatically at a disadvantage compared to off-chip passive filters, it is important to understand how to optimize their performance for a given power budget. By considering the dynamic range limitations of the G_m-C architecture in Chapter 6, we derived a figure of merit for transconductors that is useful as a design aid. With attention to the figure of merit, a survey of several types of transconductors demonstrated the merits of a new architecture that is linearized by positive feedback. Applying this approach to a 5th-order lowpass elliptical filter results in an active filter with over 60dB peak SFDR and 3.5MHz bandwidth with only 10mW of power consumption.

Combining the LNA, mixer and filter concepts, a complete GPS receiver was described in Chapter 7 that includes all of the signal path elements from LNA to bits. The receiver has a pre-limiter noise figure of only 2.8dB, resulting in an output post-correlation SNR for a noncoherent digital back end of 17dB when processing an input signal of -130dBm available source power. The total power consumption is 115mW from a single 2.5V supply.

2. RECOMMENDATIONS FOR FUTURE WORK

There are a number of issues that await exploration in future research studies. In particular, the area of CMOS modeling for RF applications is ripe for research. The behavior of the drain and gate noise coefficients (γ and δ) with respect to bias voltage and the details of MOSFET device structure is poorly understood at present. Unfortunately, the lack of a good noise model presents a substantial barrier to the implementation of CMOS receivers. A thorough experimental and theoretical investigation of noise modeling for CMOS devices would greatly reduce this barrier.

Another area that needs further investigation is the subject of substrate noise coupling in integrated radio receivers. So far, there are virtually no published reports of substantial co-integration of DSP and radio front-end components. This lack is likely due to the extreme complexities involved in modeling noise coupling mechanisms in mixed signal systems. Design techniques that can mitigate or reduce interference between digital and radio blocks are needed. A study of substrate coupling in a radio context will be necessary to enable the achievement of a true "single-chip radio".

In the future, process scaling may necessitate a reduction in supply voltage for integrated radio receivers. This trend will present a number of challenges for wide dynamic range receiver design. Thus, an investigation into low-voltage radio techniques would also be an interesting and timely topic.

Appendix A
Cross-correlation Properties of Limited Gaussian Noise Channels

Limiters are widely used in wireless receiver systems in cases where amplitude information is not essential. In the context of the GPS channel, the noise power dominates before correlation, and it is important to understand the effect of limiting on the noise statistics, particularly the effect on autocorrelation and cross-correlation. Because the former of these can be treated as a special case of the latter, this appendix analyzes the effect of limiting on the cross-correlation of two Gaussian noise processes. With this analysis as a tool, the specific case of the Weaver architecture with signal-path limiters is then considered.

1. LIMITED GAUSSIAN NOISE

Let n_1, n_2 and n_3 be independent Gaussian random variables with variance $\sigma^2 = 1$. Furthermore, define

$$x = \sqrt{c}n_1 + \sqrt{1-c}n_2 \tag{A.1}$$

$$y = \sqrt{c}n_1 + \sqrt{1-c}n_3 \tag{A.2}$$

so that

$$E\left[x^2\right] = E\left[y^2\right] = 1 \tag{A.3}$$

$$E\left[xy\right] = c \tag{A.4}$$

Hence, c is the cross-correlation coefficient of the two noise processes x and y. We can define two new processes that result from hard-limiting of x and y

$$\tilde{x} = sgn[x] \tag{A.5}$$

$$\tilde{y} = sgn[y] \tag{A.6}$$

The goal of this appendix is, then, to determine $E[\tilde{x}\tilde{y}]$.

Making use of the expressions for x and y, we can write

$$E[\tilde{x}\tilde{y}] = E\left[sgn\left(\sqrt{c}n_1 + \sqrt{1-c}n_2\right) sgn\left(\sqrt{c}n_1 + \sqrt{1-c}n_3\right)\right] \quad \text{(A.7)}$$

Using nested expectations, we can temporarily fix n_1 and expand the expression as follows,

$$E[\tilde{x}\tilde{y}] = E\left\{E\left[sgn\left(\sqrt{c}n_1 + \sqrt{1-c}n_2\right)|n_1\right] E\left[sgn\left(\sqrt{c}n_1 + \sqrt{1-c}n_3\right)|n_1\right]\right\} \quad \text{(A.8)}$$

With n_1 temporarily fixed as a constant in the two inner expectations, we can evaluate these expectations directly using the probability density function (PDF) for a Gaussian with mean $\sqrt{c}n_1$ and variance $1-c$. Thus,

$$E\left[sgn\left(\sqrt{c}n_1 + \sqrt{1-c}n_2\right)|n_1\right] = \int_{-\infty}^{\infty} \frac{sgn(u)}{\sqrt{2\pi(1-c)}} e^{-(u-\sqrt{c}n_1)^2/2(1-c)}\, du \quad \text{(A.9)}$$

We can simplify this expression through a change of variables. Let

$$v = \frac{u - \sqrt{c}n_1}{\sqrt{2(1-c)}} \quad \text{(A.10)}$$

Then, equation (A.9) becomes

$$E\left[sgn\left(\sqrt{c}n_1 + \sqrt{1-c}n_2\right)|n_1\right] = \frac{1}{\sqrt{\pi}} \int_{-\infty}^{\infty} sgn\left[\sqrt{2(1-c)}v + \sqrt{c}n_1\right] e^{-v^2}\, dv \quad \text{(A.11)}$$

The sgn function is a two-valued function that is $+1$ when its argument is positive, and -1 when its argument is negative. The argument of the sgn function in (A.10) is zero when

$$v = v_0 = -n_1\sqrt{\frac{c}{2(1-c)}} \quad \text{(A.12)}$$

With this definition of v_0, we can decompose (A.10) into two separate integrals. Thus,

$$E\left[sgn\left(\sqrt{c}n_1 + \sqrt{1-c}n_2\right)|n_1\right] = \frac{1}{\sqrt{\pi}} \int_{v_0}^{\infty} e^{-v^2}\, dv - \frac{1}{\sqrt{\pi}} \int_{-\infty}^{v_0} e^{-v^2}\, dv \quad \text{(A.13)}$$

Now, because

$$erfc(x) = 1 - erf(x) = \frac{2}{\sqrt{\pi}} \int_{x}^{\infty} e^{-v^2}\, dv \quad \text{(A.14)}$$

we can re-express (A.14) as

$$E\left[sgn\left(\sqrt{c}n_1 + \sqrt{1-c}n_2\right)|n_1\right] = \frac{erfc(v_0)}{2} - \left(1 - \frac{erfc(v_0)}{2}\right) = -erf(v_0) \tag{A.15}$$

We can use this result in (A.7) to write

$$E[\tilde{x}\tilde{y}] = E\left[erf^2(v_0)\right] \tag{A.16}$$

which expands to

$$E[\tilde{x}\tilde{y}] = \int_{-\infty}^{\infty} \frac{1}{\sqrt{2\pi}} erf^2\left[-u\sqrt{\frac{c}{2(1-c)}}\right] e^{-u^2/2} du \tag{A.17}$$

With a change of variables, we can simplify this expression somewhat. If we let $u = -\sqrt{2}v$, then

$$E[\tilde{x}\tilde{y}] = \frac{1}{\sqrt{\pi}} \int_{-\infty}^{\infty} erf^2\left[v\sqrt{\frac{c}{1-c}}\right] e^{-v^2} dv \tag{A.18}$$

In general, we have $R_{xy} = c$. Therefore, the final expression for the cross-correlation of two limited gaussian noise processes is

$$R_{\tilde{x}\tilde{y}} = E[\tilde{x}\tilde{y}] = \frac{1}{\sqrt{\pi}} \int_{-\infty}^{\infty} erf^2\left[v\sqrt{\frac{R_{xy}}{1-R_{xy}}}\right] e^{-v^2} dv \tag{A.19}$$

Equation (A.19) expresses how two Gaussian processes with a particular cross-correlation, R_{xy}, will produce limited Gaussian processes with cross-correlation, $R_{\tilde{x}\tilde{y}}$. The same expression holds for the *autocorrelation* of a *single* Gaussian process that is applied to a limiter. Figure A.1 shows a plot of equation (A.19). Clearly, limiters have a de-correlating effect on their input noise processes.

2. CROSS-CORRELATION IN THE WEAVER RECEIVER

Having determined the effect of limiters on the autocorrelation and cross-correlation properties of gaussian noise, we turn our attention to the specific case of the Weaver architecture, illustrated in Figure A.2. This architecture is exactly that presented in the historical review of Chapter 1, with the exception that limiters have been introduced in the two signal paths.

Let $n(t)$ be a Gaussian noise process with autocorrelation $R_{nn}(\Delta t)$. We can evaluate the cross-correlation at point A in Figure A.2 as follows

$$R_{IQ}(t,s) = E\left[n(t)cos(\omega t)n(s)sin(\omega s)\right] \tag{A.20}$$

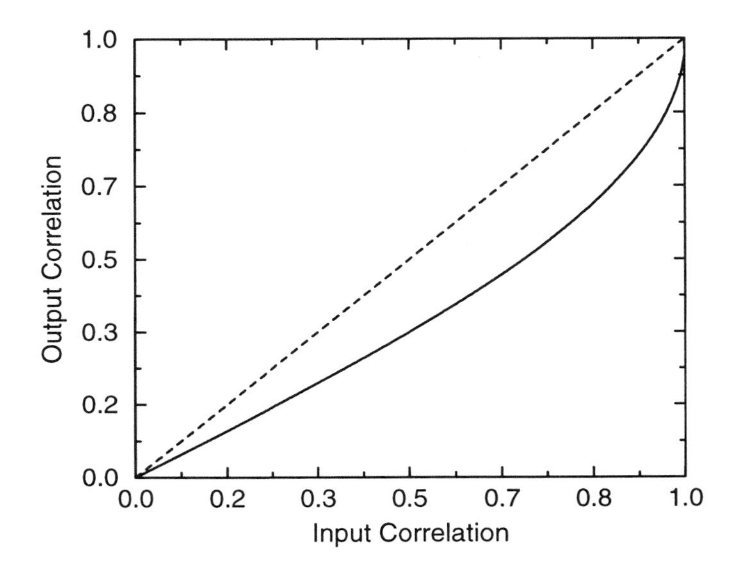

Figure A.1. The effect of a limiter on the cross-correlation or autocorrelation of a Gaussian noise process.

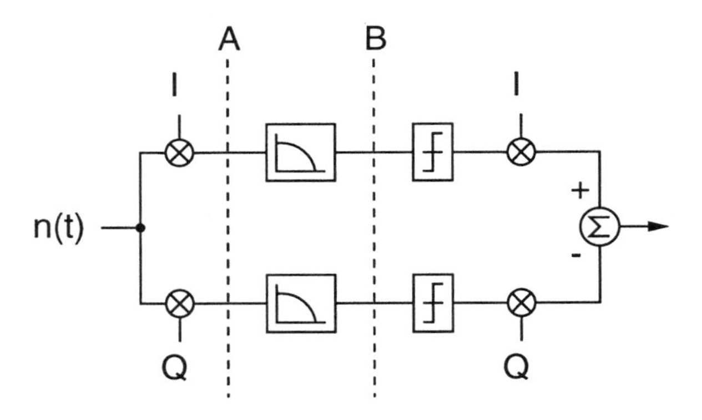

Figure A.2. Simplified block diagram of the CMOS GPS receiver, including a coherent back-end demodulation to baseband.

which simplifies to

$$R_{IQ}(t,s) = \frac{R_{nn}(\Delta t)}{2}\left[sin(\omega s + \omega t) + sin(\omega \Delta t)\right] \tag{A.21}$$

So that if $R_{nn}(\Delta t) = \sigma_n^2 \delta(\Delta t)$ then

$$R_{IQ}(t) = \frac{\sigma_n^2}{2} sin(2\omega t) \tag{A.22}$$

where $\Delta t = s - t$. Hence, if $n(t)$ is a white noise process, the cross-correlation between the two channels alternates sign at twice the LO frequency.

Similarly, we can calculate the cross-correlation at point B, after the low-pass filter. If the two filters are matched and each has an impulse response $h(t)$, then $R_{IQ}(t, s) =$

$$E\left\{[(n(t)cos(\omega t)) * h(t)]\,[(n(s)sin(\omega s)) * h(s)]\right\} \tag{A.23}$$

which can be expanded to

$$\int_{\infty}^{\infty}\int_{\infty}^{\infty} cos(\omega x) sin(\omega y) h(t-x) h(s-y) R_{nn}(x-y)\, dx\, dy \tag{A.24}$$

Now, if we assume that $R_{nn}(\Delta t) = \sigma_n^2 \delta(\Delta t)$ and if we restrict ourselves to the case where $s = t$, then after some simplifications

$$R_{IQ}(t) = \frac{1}{2} sin(2\omega t) * h^2(t) \tag{A.25}$$

From this expression, if $h(t)$ is a low-pass filter, and if ω is the local oscillator frequency and is much greater than the cutoff frequency of the filter, then $R_{IQ}(t) = 0$. Thus, the two channels are de-correlated by the action of the low-pass filters.

Note that complete decorrelation does *not* occur in general, but rather depends on the assumption of a white input noise process. More generally, it is sufficient for the image and desired sidebands to have equal noise powers and to be white only within the bandwidth of the filter, $h(t)$. These constraints are satisfied in a low-IF architecture, due to the narrow filter bandwidth and the proximity of the desired and image frequencies.

Because the two channels are uncorrelated before quantization, they remain so after quantization so that no distortion of the cross-correlation occurs. Subsequent summation of the I and Q channels therefore leads to a 3-dB SNR improvement, which is precisely the result expected if the image noise were cancelled.

Appendix B
Classical MOSFET Noise Analysis

This appendix presents the classical approach to formulating the minimum noise figure of a MOSFET device. The presentation here draws heavily on the work in [43], to which the reader is referred for a complete treatment of the classical theory.

Many texts on CMOS amplifier design indicate that the MOSFET device has a single dominant noise source: channel thermal noise. This source is commonly modeled as shown in Figure B.1(a) and has a power spectral density of

$$\overline{i_d^2} = 4kTB\gamma g_{do}. \tag{B.1}$$

The MOSFET also has an additional source of noise arising from capacitive coupling to the noisy channel that causes a noise current to flow in the gate. This *induced gate current noise* is easily modeled as a shunt current source in the gate circuit of the device, as shown in Figure B.1(a), and has a power

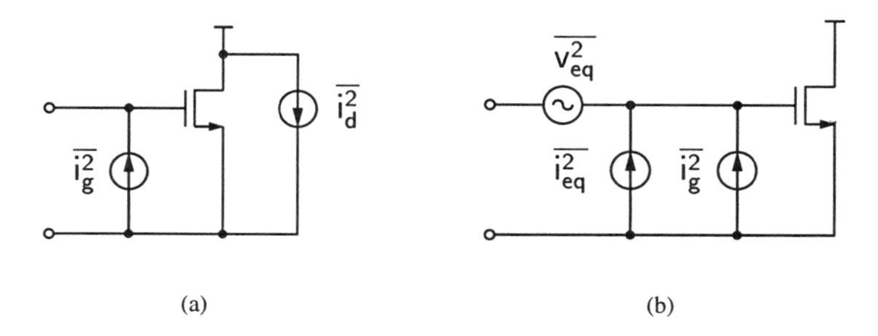

Figure B.1. Equivalent noise models for a MOSFET device with drain and gate current noise. (a) Physical model. (b) Equivalent model with input-referred sources.

spectral density of

$$\overline{i_g^2} = 4kTB\delta g_g \tag{B.2}$$

$$g_g = \frac{\omega_0^2 C_{gs}^2}{5g_{d0}}. \tag{B.3}$$

Note that this source of noise is not a white noise source. It has an increasing power spectral density proportional to ω_0^2 and is perhaps better described as "blue", to retain the optical analogy. In addition, the induced gate noise is *partially correlated* with the drain noise. To account for this correlation, the gate current noise is divided into two terms: i_g' (which is *fully correlated* with the drain current noise), and i_g'' (which is *uncorrelated* with the drain current noise). The degree of correlation is then conveniently expressed with a correlation coefficient, c, defined as

$$c = \frac{\overline{i_g i_d^*}}{\left(\overline{i_g^2}\,\overline{i_d^2}\right)^{\frac{1}{2}}}. \tag{B.4}$$

Using c, we express i_g as

$$\overline{i_g^2} = \overline{\left(i_g' + i_g''\right)^2} = 4kTB\delta g_g |c|^2 + 4kTB\delta g_g \left(1 - |c|^2\right). \tag{B.5}$$

The noise figure can be evaluated using equations from [43] if the drain current noise is referred to the input, as shown in Figure B.1(b). A simple analysis shows that

$$\overline{v_{eq}^2} = \frac{\overline{i_d^2}}{g_m^2} = \frac{4kTB\gamma g_{d0}}{g_m^2} \tag{B.6}$$

and

$$\overline{i_{eq}^2} = \frac{\overline{i_d^2}\omega_0^2 C_{gs}^2}{g_m^2} = \frac{4kTB\gamma g_{d0}\omega_0^2 C_{gs}^2}{g_m^2} = \overline{v_{eq}^2}\left(\omega_0 C_{gs}\right)^2 \tag{B.7}$$

where g_{d0} is the zero-bias drain conductance of the MOSFET device. Because $\overline{v_{eq}^2}$ and $\overline{i_{eq}^2}$ are derived from network transformations on the same noise source, they are *fully correlated.*

We can now evaluate the minimum noise figure of the MOSFET device, including the effects of induced gate noise. The most difficult part, mathematically, is the evaluation of the correlation admittance, Y_c, which relates the correlated portions of the total equivalent input noise current and voltage:

$$Y_c = \frac{i_{eq} + i_g'}{v_{eq}} = sC_{gs} + g_m \frac{i_g'}{i_d}. \tag{B.8}$$

Multiplying the second term's numerator and denominator by i_d^* and averaging,

$$Y_c = sC_{gs} + g_m \frac{\overline{i'_g i_d^*}}{\overline{i_d^2}} = sC_{gs} + g_m \frac{\overline{i_g i_d^*}}{\overline{i_d^2}}. \tag{B.9}$$

The second step in Equation (B.9) is true because i''_g is uncorrelated with i_d, and hence does not contribute to the cross-correlation term. We can re-express (B.9) as follows

$$Y_c = sC_{gs} + g_m \frac{\overline{i_g i_d^*}}{\left(\overline{i_g^2}\;\overline{i_d^2}\right)^{\frac{1}{2}}} \left(\frac{\overline{i_g^2}}{\overline{i_d^2}}\right)^{\frac{1}{2}}. \tag{B.10}$$

We now can recognize the definition of c from Equation (B.4). Substituting expressions for $\overline{i_g^2}$ and $\overline{i_d^2}$ from Equations (B.3) and (B.1), respectively, we find that

$$Y_c = sC_{gs} + g_m c \sqrt{\frac{\delta \omega_0^2 C_{gs}^2}{5\gamma g_{d0}^2}} \tag{B.11}$$

An analysis by van der Ziel [1] has shown that the correlation coefficient, c, is imaginary, and has a value of $j0.395$. Using this fact, (B.11) can be simplified.

$$Y_c = G_c + jB_c = j\omega_0 C_{gs} \left[1 + \alpha |c| \sqrt{\frac{\delta}{5\gamma}}\right] \tag{B.12}$$

where

$$\alpha = \frac{g_m}{g_{d0}}. \tag{B.13}$$

Note that $Y_c \neq sC_{gs}$. This inequality means that the source impedance required for maximum power transfer (a conjugate impedance match where $B_s = -sC_{gs}$) is *different* than that which is required for minimum noise figure ($B_s = -B_c$). Thus, one cannot have a perfect input match and minimum noise figure simultaneously.

We can define a term called the *equivalent noise resistance* that describes the equivalent noise voltage power in terms of a resistor. It is defined to be

$$R_n = \frac{\overline{v_{eq}^2}}{4kTB} = \frac{\gamma g_{d0}}{g_m^2}. \tag{B.14}$$

Similarly, we define the *equivalent uncorrelated noise conductance* to describe the equivalent noise power of the *uncorrelated* portion of the gate current noise.

Hence,

$$G_u = \frac{\overline{i_g{''}^2}}{4kTB} = \frac{\delta\omega_0^2 C_{gs}^2 \left(1 - |c|^2\right)}{5g_{d0}}. \tag{B.15}$$

With these definitions, we can directly apply the theory outlined in [43] to determine the minimum noise figure. The optimum source admittance is

$$Y_{opt} = G_{opt} + jB_{opt} \tag{B.16}$$

where

$$G_{opt} = \sqrt{G_c^2 + \frac{G_u}{R_n}} \tag{B.17}$$

$$B_{opt} = -B_c. \tag{B.18}$$

In this case, substitution of R_n, G_c, G_u and B_c results in

$$Y_{opt} = G_{opt} + jB_{opt} = \omega_0 C_{gs}\sqrt{\frac{\delta\alpha^2}{5\gamma}\left(1 - |c|^2\right)} - j\omega_0 C_{gs}\left[1 + \alpha|c|\sqrt{\frac{\delta}{5\gamma}}\right]. \tag{B.19}$$

With this optimum source admittance, the corresponding minimum noise figure is

$$F_{min} = 1 + 2R_n\left(G_{opt} + G_c\right) \tag{B.20}$$

$$= 1 + \sqrt{\frac{4}{5}\delta\gamma}\left(\frac{\omega_0}{\omega_T}\right)\sqrt{(1 - |c|^2)} \tag{B.21}$$

$$\geq 1 + 0.773\left(\frac{\omega_0}{\omega_T}\right) \tag{B.22}$$

where the coefficient 0.773 is only valid in the long-channel limit.

A few observations are in order. Note that, when induced gate noise is included, F_{min} is well-defined, and $F_{min} > 1$ (0dB). Secondly, the optimum source admittance is a parallel R-L circuit. As a result, we can define an optimum Q which corresponds to the optimum source admittance. Hence,

$$Q_{opt} = \left|\frac{B_{opt}}{G_{opt}}\right| = \frac{\sqrt{\frac{5\gamma}{\delta\alpha^2}} + |c|}{\sqrt{1 - |c|^2}} \approx 2.162. \tag{B.23}$$

Note that this optimum Q is very similar to Q_{L,opt,G_m}. Because (B.23) only depends on the *ratio* of γ and δ, it is reasonable to expect that short channel phenomena will have only a second-order effect on Q_{opt}.

Appendix C
Experimental CMOS Low-Noise Amplifiers

This appendix presents the results of two experiments with low-noise amplifiers in CMOS. The first experiment is with a single-ended design in a 0.6μm process, and the second is with a differential design in a 0.35μm process.

1. AN EXPERIMENTAL SINGLE-ENDED LNA

To probe further the ability of CMOS to deliver low noise amplification at 1.57542GHz, we have implemented a LNA in a 0.6μm CMOS technology provided through the MOSIS service (0.35μm L_{eff}). The only information about the technology available at the time of design referred to interlayer dielectric thicknesses, sheet resistances, and diffusion capacitances. Thankfully, the value of t_{ox} was also available, making possible a crude extrapolation from 0.8μm models to provide some basis for simulation. The success of the implementation demonstrates that knowledge of device capacitances is the most important factor in the design of tuned amplifiers.

1.1 IMPLEMENTATION

The width of the input device was initially chosen without regard to the induced gate noise term because the detailed nature of gate noise was unknown to the authors at design time. It will prove useful to know the optimum width for this technology so that we can determine whether our performance is limited by the induced gate noise or by the drain current noise. From Figure 4.8, the optimum Q_L for a power dissipation of 7.5mW (which corresponds to the measured P_D of the first stage of our LNA) is about 4.5, with a corresponding F_{min,P_D} of 2.1dB. We can immediately determine the optimum width to be

$$W_{M_1,opt,P_D} = \left[\frac{2}{3}\omega_0 L C_{ox} R_s Q_{L,opt,P_D}\right]^{-1} \approx 496\mu m \qquad \text{(C.1)}$$

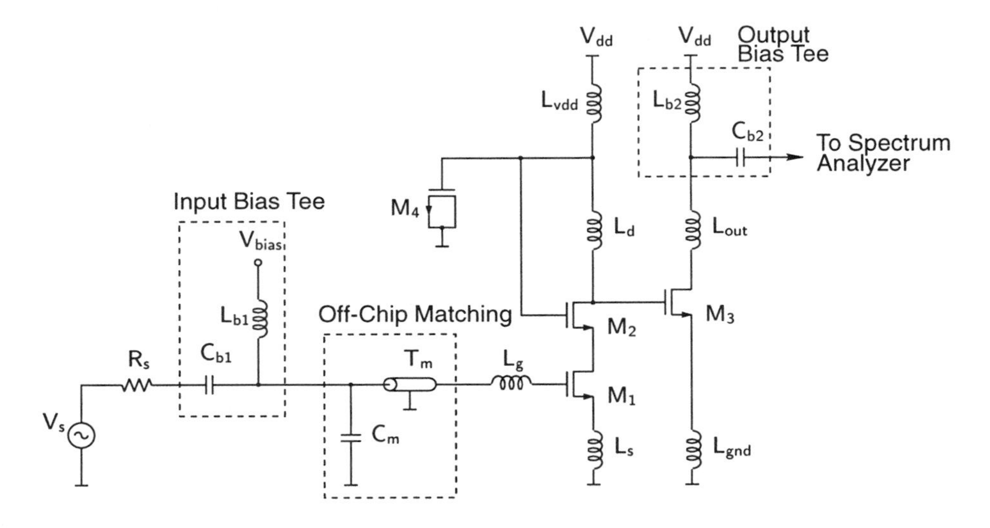

Figure C.1. Complete schematic of the LNA, including off-chip elements.

where $\omega_0 = 10$Grps, $L = 0.35\mu m$, $C_{ox} = 3.84mF/m^2$, and $R_s = 50\Omega$. The actual width of M_1, as implemented, is about 403μm, which corresponds to a Q_L of 5.5, still very close to the minimum noise figure point for 7.5mW of power dissipation. Because our Q_L is greater than the optimum, we expect that our measured performance will be limited by the gate noise. Note, however, that the predicted F neglects any contribution to the noise figure by parasitic losses, particularly those due to on-chip spiral inductors, which influence the noise figure of the LNA. Accordingly, the amplifier will possess a noise figure which is greater than 2.1dB.

The complete schematic of the LNA is shown in Figure C.1. The amplifier is a two-stage, cascoded architecture. The drain of M2 is tuned by a 7nH on-chip spiral inductor, Ld. This inductor resonates with the total capacitance at the drain of M2, including C_{gs} of M3. Transistor M3 serves as an open-drain output driver providing 4.6dB of gain, and the amplifier uses the test instrument itself as the load. Note that M3 has a gate width of about 200μm, or half of M1.

Four of the inductors shown (Ls, Lgnd, Lvdd and Lout) are formed by bondwire inductances. Of these four, Ls is the only one whose specific value is significant in the operation of the amplifier, since it sets the input impedance of the LNA. Lgnd and Lout are unwanted parasitics, so their values are minimized by proper die bonding. Lvdd aids in supply filtering with M4, which acts as a supply bypass capacitor. Because a large value of inductance is beneficial for this use, Lvdd is formed from a relatively long bondwire.

Figure C.2. Die photo of the LNA.

Due to the lack of simulation models before fabrication, a flexible topology was chosen which would permit post-fabrication adjustment of the bias points of M1 and M2. The input matching is accomplished with the aid of an off-chip network. Off-chip tuning was required because the necessary value of Lg was prohibitively large for on-chip fabrication. However, a 4nH inductor was integrated on-chip in series with the gate of M1. This inductor, together with the input bondwire inductance, reduces the matching burden of the off-chip network. Unfortunately, it also introduces additional resistive losses which degrade the noise performance of the LNA.

A die photo of the LNA is shown in Figure C.2. The two spiral inductors are clearly visible. The input pad is on the lower left corner of the die. The spiral on the left is a 4nH inductor which forms a portion of Lg. The spiral on the right is a 7nH inductor that tunes the output of the first stage. The spirals are fabricated in metal-3, which permits Q's of about 3 to be achieved. This value of Q is typical of on-chip spiral inductors that have been reported in the literature [51]. To improve the Q slightly, the inductors are tapered so that the outer spirals use wider metal lines than the inner spirals. The goal of this tapering is to distribute the loss to yield a roughly constant loss per turn. A magnetic field solver, FastHenry, was used during the design of the LNA to predict the values of inductance and the winding loss associated with various geometries. From these simulations, we determined that tapering provides a slight, but welcome, increase in Q (approximately 20%).

1.2 EXPERIMENTAL RESULTS

To test the LNA, the die was mounted in a high-frequency package and bonded. The measured gain (S21) of the amplifier appears in Figure C.3. The

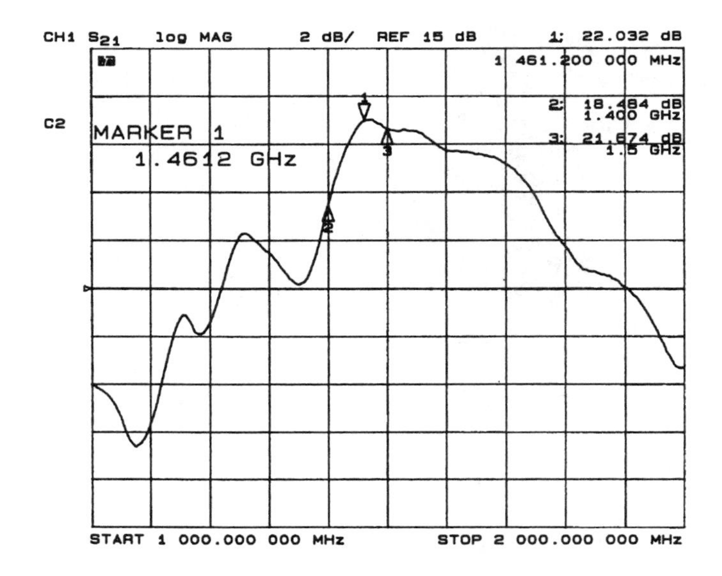

Figure C.3. Measured S21 of the LNA

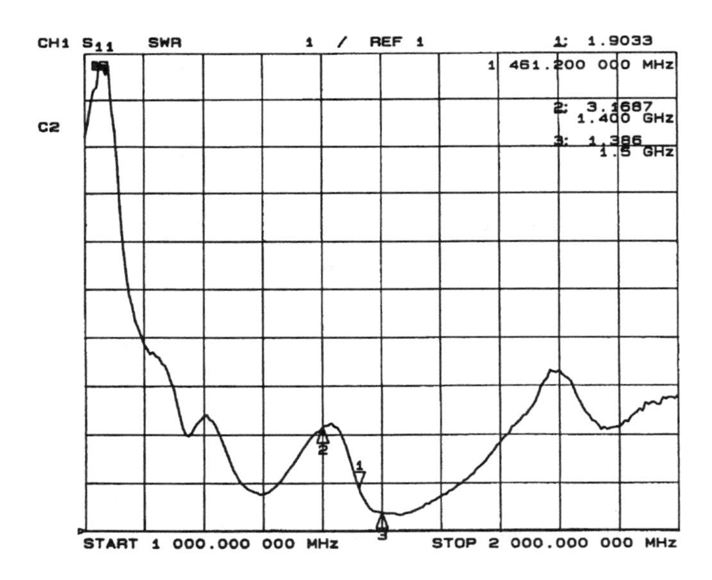

Figure C.4. Measured S11 of the LNA

gain has a peak value of 22dB at 1.46GHz and remains above 20dB to almost 1.6GHz. The bandpass nature of the amplifier is evident from the plot. The input reflection coefficient (S11) is also plotted in Figure C.4. The input

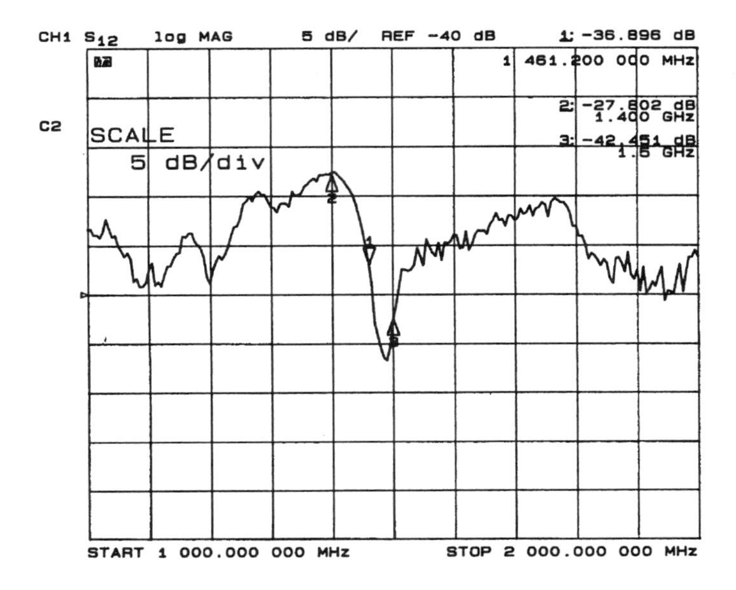

Figure C.5. Measured S12 of the LNA

VSWR at 1.5GHz is quite good (about 1.4) with the addition of off-chip tuning elements.

It is interesting that both plots exhibit some anomalies at about 1.4GHz. On the S21 curve, the gain begins to dip sharply, whereas the S11 plot shows a bump in the reflection coefficient. This point is indicated by marker 2 on both plots. An examination of the reverse gain of the amplifier (S12) in Figure C.5 provides a plausible explanation for these anomalies. Marker 2 is positioned at the same frequency as in the two previous plots. Note that it coincides with a pronounced peak in the reverse gain. Indeed, the approximate loop gain magnitude of the LNA at marker 2 is -6dB. This value is insufficient to cause oscillation of the amplifier, but is nonetheless substantial. Accordingly, we are compelled to attribute the formerly mentioned anomalies to this reverse isolation problem.

Another feature of the S12 characteristic is a sharp null at 1.5GHz. This null is a clue to the source of our troubles. In Figure C.6, a partial schematic of the LNA is shown along with various significant parasitic capacitances. The substrate of the die was connected to the lowest inductance signal ground, Lgnd. As shown in the diagram, this choice degrades the reverse isolation by allowing signal currents in the output driver to couple back to the input through the large parasitic capacitance of the gate inductance and its bond pad. There are actually two significant paths for this to occur, opening the possibility of cancellation at a particular frequency. Indeed, a significant phase shift along

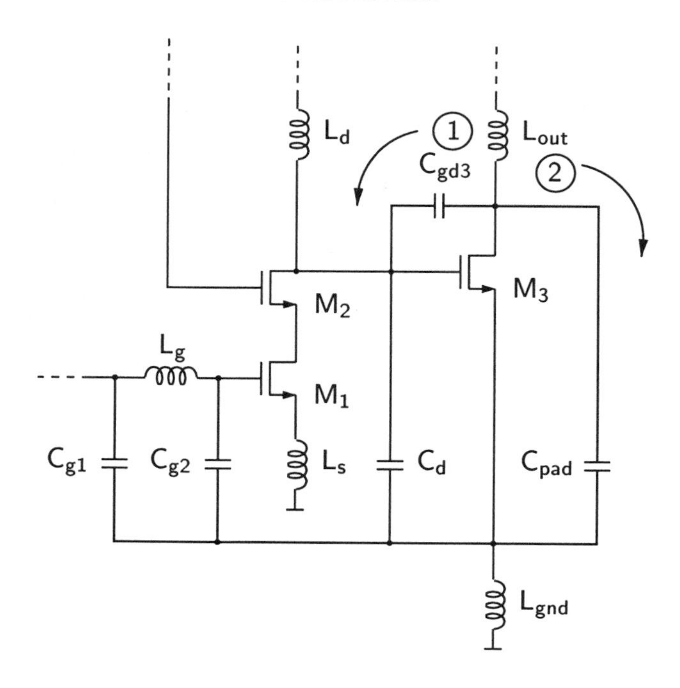

Figure C.6. Detailed LNA schematic showing parasitic reverse paths.

path 1 in the diagram occurs near the resonance of Ld and Cd. A null in the reverse gain could thus occur near this frequency. This problem could be mitigated by terminating the substrate differently, or by moving to a differential structure.

The noise figure and gain of the LNA are plotted in Figure C.7. From this plot, we can see that at Vdd=1.5V, the LNA exhibits a 3.5dB noise figure with 22dB of forward gain. The power dissipation is 30mW total. Of this power, only 7.5mW is consumed by the first amplifier stage. The other 22.5mW is used to drive 50Ω with the open-drain output driver. This added power could be nearly eliminated if the LNA were to drive an on-chip mixer rather than an off-chip transmission line.

Although the measured noise figure exceeds the theoretical minimum of 2.1dB, it is a simple matter to account for the difference. In particular, our theoretical predictions must be modified to include the loss of the 4nH spiral inductor, which contributes significantly to the noise figure, and to account for the actual impedance level at the LNA input, as determined by $\omega_T L_s$. In the final amplifier, $\omega_T L_s$ was less than 50Ω. In fact, the real portion of the input impedance, before matching, was about 35Ω. If we assume that the 4nH inductor possesses a Q of about three, then it would contribute about 0.38 to

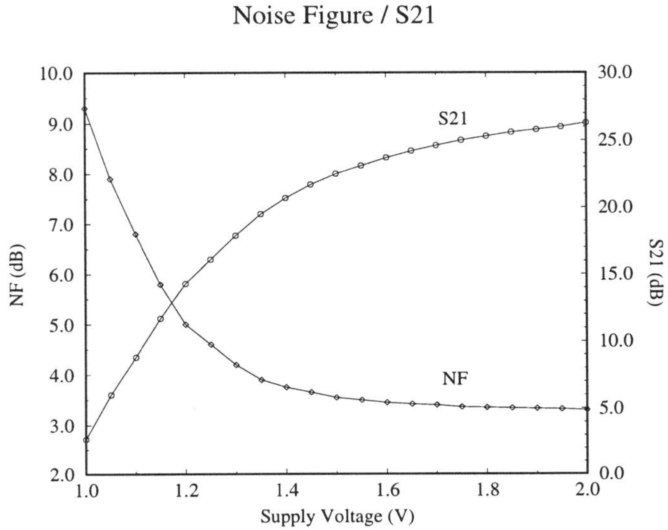

Figure C.7. Noise figure and forward gain of the LNA.

F in a 35Ω environment. In addition, the theoretical minimum increases to about 2.5dB when Rs is 35Ω. These two effects therefore elevate the predicted noise figure from 2.1dB to 3.3dB. The remaining 0.2dB may be attributed to the second stage of the amplifier.

A two-tone IP3 measurement was performed on the LNA and the results are shown in Figure C.8. The two tones were applied with equal power levels at 1.49GHz and 1.5GHz. The measurement indicates a -9.3dBm input-referred third-order intercept point (+12.7dBm output-referred). The linearity is primarily limited by M3, due to the gain which precedes it.

The measured performance of the LNA is summarized in Table C.1.

2. AN EXPERIMENTAL DIFFERENTIAL LNA

As an alternative to a single-ended LNA, one might select a differential architecture. As demonstrated in the single-ended LNA experiment, the influence of substrate parasitics can be pernicious. Differential architectures will be somewhat immune to such common-mode impedances due to the differential symmetry. In addition, in the context of a complete integrated receiver system, rejection of common-mode noise is important. Because the LNA is the most sensitive signal block in the receive path, it is important to minimize coupling between other blocks and the LNA input, for stability reasons as well as supply and substrate noise reasons.

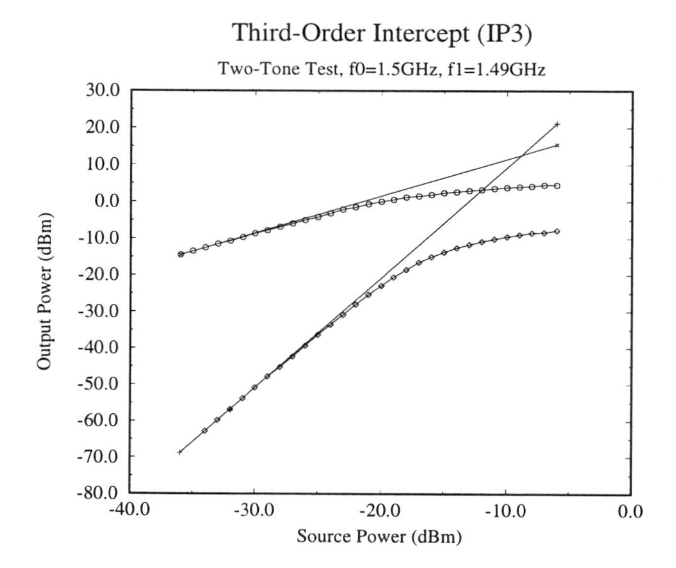

Figure C.8. Results of two-tone IP3 measurement.

Table C.1. Single-ended LNA Performance Summary

Frequency	1.5GHz
Noise Figure	3.5dB
S21	22dB
IP3 (Output)	12.7dBm
1dB Compression (Output)	0dBm
Supply Voltage	1.5V
Power Dissipation	30mW
(First Stage)	7.5mW
Technology	0.6μm CMOS
Die Area	0.12mm^2

The penalty for selecting a differential architecture, however, is that twice the power must be consumed to achieve the same noise performance. Furthermore, an off-chip balun will be required for interfacing to a single-ended RF filter and antenna, and this balun will introduce loss, thereby degrading the noise figure of the system. For modern ceramic hybrid baluns in this frequency range, a loss of about 0.5dB can be expected. In some applications, the balun can be eliminated through the use of a balanced antenna and RF filter.

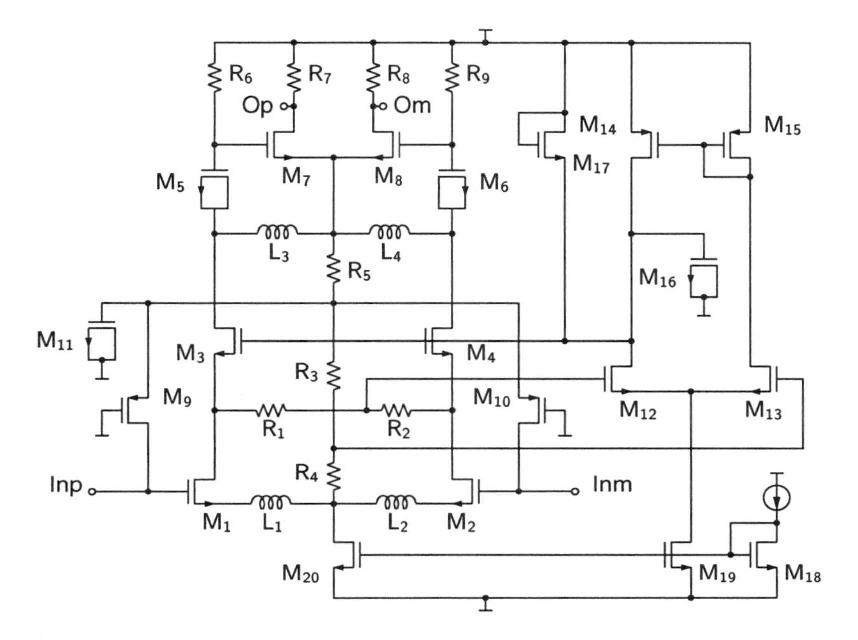

Figure C.9. Differential LNA circuit diagram

2.1 IMPLEMENTATION

Figure C.9 shows a circuit level description of the differential LNA. It consists of two stages: the input stage, formed by transistors M1 through M4, and the output stage, formed by transistors M7 and M8. The input stage is cascoded for a number of reasons. The first is to reduce the influence of the gate to drain overlap capacitance, C_{gd}, on the LNA's input impedance. Specifically, the Miller effect tends to lower the input impedance substantially, complicating the task of matching to the input. In addition to mitigating the Miller effect, the use of a cascode improves the LNA's reverse isolation, which is important in the present application for suppressing local oscillator (LO) feedthrough from the mixer back to the LNA's radio frequency (RF) input. Furthermore, because the output of the first stage is tuned with spiral inductors, L3 and L4, the LNA's stability might be compromised without the cascode, due to interaction between the inductive load and the input matching network through C_{gd}. It should be noted, however, that a noise penalty is incurred when using a cascode. But, with proper attention to the layout of the devices, the additional noise can be minimized. For input matching purposes, the LNA uses inductive source degeneration via two on-chip spirals, L1 and L2.

The bias current of the input stage is reused in the output stage, decreasing the power by a factor of two. The low threshold voltage of this process permits four devices to be stacked, provided that adequate bias control is included. The

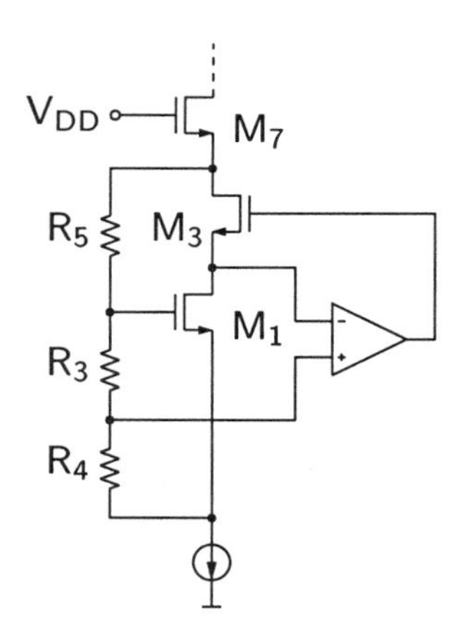

Figure C.10. Single-ended version of the DC biasing technique

goal is to use the minimum V_{ds} to keep devices M1 and M2 in saturation, while leaving some room for signal swing. This will also, hopefully, benefit the noise performance by limiting the peak electric field in the input devices.

Figure C.10 illustrates the active common-mode feedback technique that permits the amplifier to operate reliably on a 1.5V supply, independent of process, supply, and temperature variations. Resistors R3 and R4 sense a fraction of the input devices' common-mode V_{gs} level. This fraction becomes the reference to which the input devices' common-mode V_{ds} level is servoed. An operational amplifier, formed by transistors M12 through M15, is used to close the biasing loop, with adjustments to the input devices' common-mode V_{ds} level being made via the gate voltage on the cascode devices. Resistor R5 permits extra headroom at the drains of M3 and M4, since the signal swing at these nodes can be large.

The implemented width is only 290μm, because the detailed nature of the gate noise was unknown to the authors when this amplifier was designed. However, the noise figure curve has a broad minimum, so the achievable noise figure is little affected by using transistors of this width, at least in principle.

2.2 EXPERIMENTAL RESULTS

The LNA was integrated in a 0.35μm CMOS technology with only two metal layers. A die photograph is shown in Figure C.11. The aspect ratio of the silicon is somewhat unusual because this project was designed to fit in the scribe lane of a wafer that was primarily devoted to other dice. Accordingly, the dimensions are 350μm x 2.4mm. The results of experimental measurements are summarized in Table C.2 and discussed in detail below.

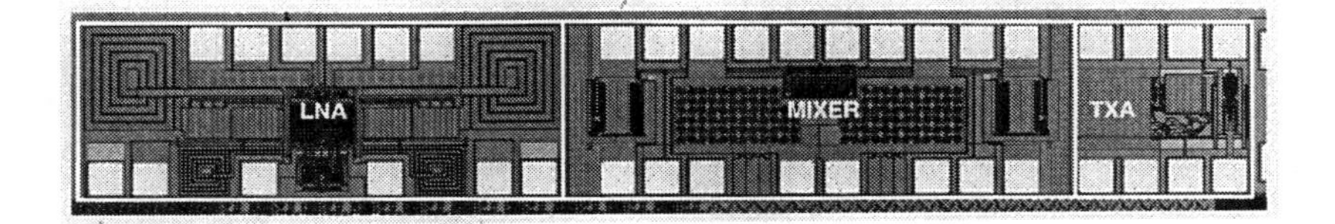

Figure C.11. Die photo

Table C.2. Differential LNA Performance Summary

Frequency	1.57542GHz
Noise Figure	3.8dB
S21	17.0dB
S12	$\leq$ -52dB
IP3 (Input)	-6dBm
1dB Compression (Input)	-20dBm
Power Dissipation	12mW
Supply Voltage	1.5V
Technology	0.35μm CMOS
Die Area	0.84mm^2

The test board for the LNA used a low-loss dielectric, and contained auxiliary test structures, to permit measurement of the insertion loss of board traces, baluns, and connectors. As a result, the noise figure of the LNA could be measured with a precision of $\sim \pm 0.2$dB.

The theoretical noise figure for this amplifier is plotted as a function of device width in Figure C.12. The theoretical noise figure for power-matched devices with a 100-Ω impedance level is shown by the lower curve, with a predicted noise figure of 1.8dB. The measured noise figure diverges from this number substantially. In part, this difference is due to the fact that the complete amplifier has more than one noise contributor; however this is not sufficient to account for the discrepancy.

A measurement of the input impedance of the LNA revealed the primary reason for the difference. The real part of the input impedance was found to be only 40Ω differential, rather than the desired 100Ω differential. This gross difference is partially due to the influence of the overlap capacitance of the input devices, which lowers the impedance seen at the gates of those devices. This behavior was observed in simulations of the LNA's input impedance, but unfortunately, the impact of the reduced impedance on the noise figure was not fully appreciated at the time. Furthermore, increasing the inductance of the

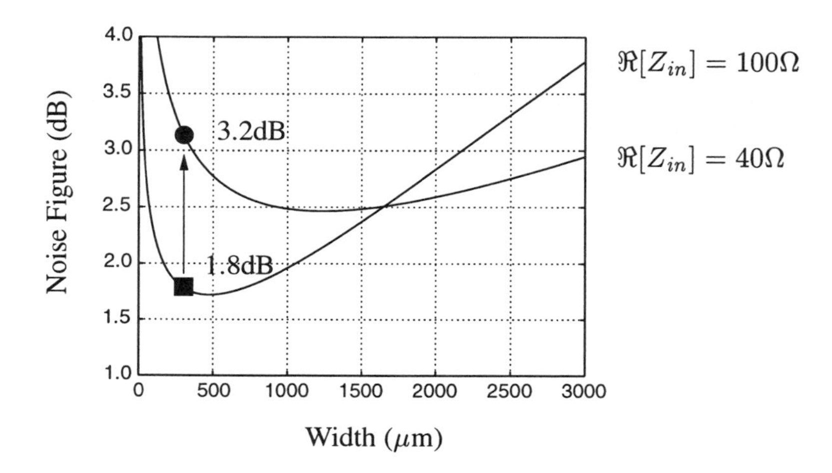

Figure C.12. Noise figure vs. device width for $R_{in} = 100\Omega$ and $R_{in} = 40\Omega$

source spiral inductors, L1 and L2, which is necessary to combat this effect, would reduce the LNA's gain, leading to an increase in the mixer's relative noise contribution.

The noise figure curve of the earlier section can be re-plotted in light of this information. Because we are matching to a lower impedance, one might expect the noise contribution of the input devices to be more significant relative to this reduced impedance. Indeed, this is the case, as is evident in the plot of Figure C.12. Both noise figure plots represent predictions for the performance of an isolated device of the stated width, assuming 12mW power consumption in the final amplifier. As can be seen, the chosen width of 290μm is substantially removed from the optimum point on the 40Ω curve. Also, the optimum point on this new curve is itself 0.7dB higher than on the 100Ω curve. These compounding effects illustrate the penalty in undershooting the desired input impedance.

The revised prediction anticipates a 3.2dB noise figure from the input pair alone. Thus, the observed total noise figure of 3.8dB is reasonable, given that other devices in the circuit contribute noise in a second-order fashion. For example, the cascode devices, the load inductors, and the output stage transistors all have noise, which contributes some small amount to the noise figure. The forward gain (S21) and noise figure are plotted in Figure C.13.

One salient feature of the differential LNA architecture that should be mentioned is its reverse gain (S12), which was measured to be less than -52dB between 1GHz and 2GHz. Good reverse isolation is required to attenuate local oscillator leakage from the mixer back to the RF input of the LNA. The use of a cascode structure in the LNA's input stage helps to reduce reverse feedthrough,

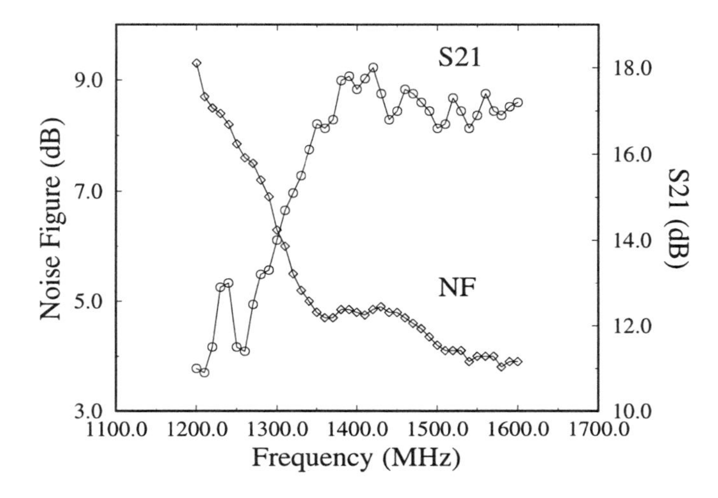

Figure C.13. LNA noise figure/S21 measurement

and this good reverse isolation is augmented by the fact that the substrate appears as an incremental ground, to first order, for differential signals.

Appendix D
Measurement Techniques

This appendix describes in detail the experimental setups that are used to gather the data presented in Chapter 7. The first section presents the noise figure measurement technique for the low-noise amplifier. The next section presents techniques for determining the noise figure, linearity and frequency response of the receiver, as measured at a test point just before the limiting amplifier chain. Finally, the last section describes how the entire receiver is tested with a pseudo-noise modulated carrier to emulate the GPS signal.

1. LNA NOISE FIGURE MEASUREMENTS

To measure the noise figure of the low-noise amplifier in the GPS receiver, a separate LNA test structure is implemented. The only difference between the test structure and the LNA in the receiver itself is that the test structure has an open-drain output driver for driving the 50Ω test instrument. Because it drives the instrument directly, the test LNA has less available power gain than the receiver LNA. However, the available power gain is sufficiently high (15dB) for noise measurement purposes.

Figure D.1 shows the experimental setup used for LNA noise figure measurements. The four layer test board is implemented on a low-loss dielectric material called RO-4003, available from Rogers Corp. A thin, 8-mil dielectric reduces the conductor width necessary for a 50Ω microstripline to only 15 mills.

A key part of this setup is the input matching network and balun. The balun is a ceramic hybrid balun available from Murata. Its insertion loss is of primary importance in noise measurements because this loss adds directly to the measured noise figure. The Murata balun provides about 0.5dB of insertion loss, typically. To calibrate out the losses preceding the LNA, a special test structure is built on the same board with two back-to-back connector/balun assemblies.

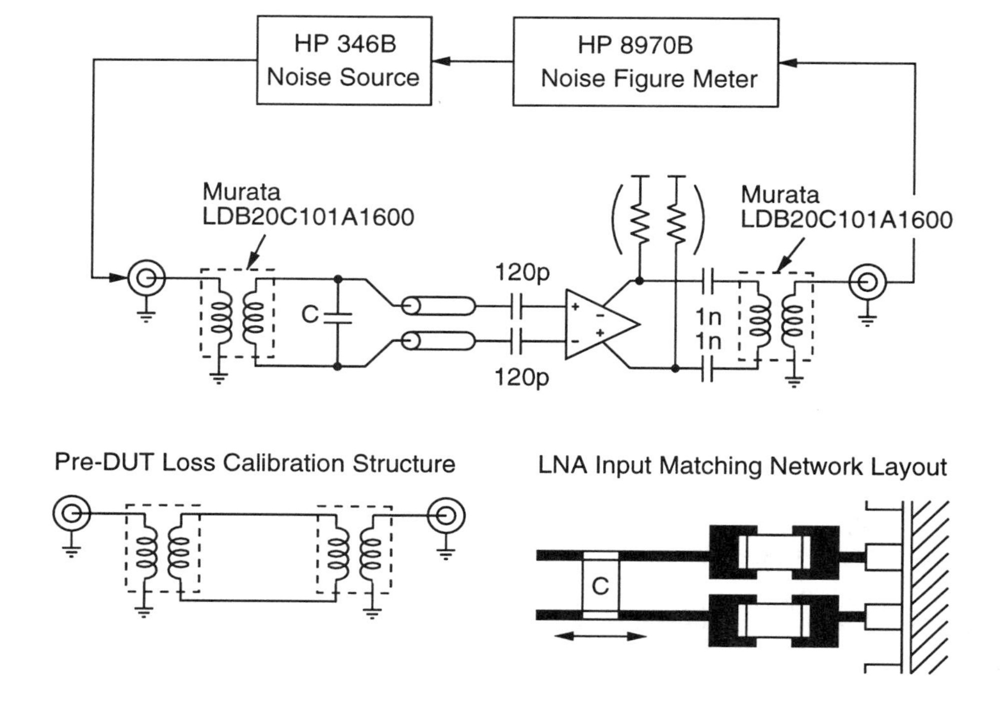

Figure D.1. Experimental setup for LNA noise figure measurements.

This calibration assembly is critical for making accurate measurements because pre-DUT insertion losses can easily exceed 1dB.

The input matching network consists of two transmission lines and a single SMD capacitor bridging the two lines. As shown in Figure D.1, the position and value of the matching capacitor can be adjusted to achieve the desired input match. This technique greatly simplifies the task of input matching. The capacitor can be held with a pair of tweezers and placed at various points along the line while S11 measurements are performed. Thus, refinement of the match proceeds easily. Once the optimum value and position are determined, the capacitor is soldered in place. It is found experimentally that matching with this technique can be done in 10 minutes, typically, compared to an hour or more for matching with a lumped component L-match.

The output of the LNA drives another Murata balun, though a connector to the HP8970B noise figure meter. For LNAs with open-drain output drivers, a simple bias tee can be implemented on the board with a pull-up inductor or resistor and a 1nF coupling capacitor. With this setup, measurement accuracy on the order of 0.2dB can be easily obtained. To achieve this accuracy, it is important that the output of the LNA be impedance-matched to the input

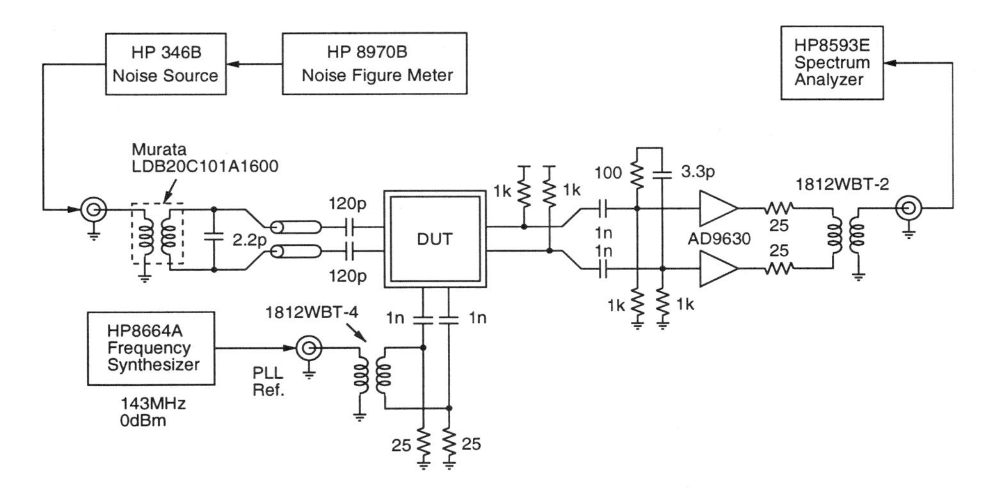

Figure D.2. Experimental setup for receiver noise figure measurements.

of the noise figure meter. This requirement is due to the fact that the meter assumes that the insertion gain of the LNA is identical to its available gain, and this is only true when the output of the LNA is matched. The insertion gain measurement is used by the meter to correct for its own contribution to the measured noise figure. Thus, any difference between the insertion gain and the available gain leads to an error.

2. PRE-LIMITER RECEIVER MEASUREMENTS

To verify the receiver signal path, several measurements are taken at a test point just preceding the limiting amplifier chain. These measurements include a noise figure measurement, an IP3 measurement and a signal path frequency response measurement.

Figure D.2 shows the experimental setup for the system noise figure measurement. The primary complication with noise figure measurements of the entire signal path is that the signal path bandwidth is only 3.5MHz, whereas the input bandwidth of the noise figure meter is 10MHz. In addition, the IF frequency is below the minimum measurement frequency of the HP8970B meter. So, the measurement must be performed with a spectrum analyzer instead of the noise figure meter.

The LNA input matching circuitry is identical to that presented in Figure D.1. An off-chip frequency reference is supplied with the HP8664A frequency synthesizer. This reference is used by the on-chip PLL for synthesis of the 1.573GHz local oscillator. Note that the reference is brought on-chip as a low-swing differential clock to reduce coupling to the substrate.

The test point output is an open drain amplifier that is very similar to the intermediate frequency amplifier. The output signal is ac coupled to two AD9630 high-speed low-distortion buffers that provide sufficient transducer power gain into 50Ω to make a meaningful noise measurement. The noise of the buffers contributes negligibly to the total noise figure. The buffered differential signal is transformed into a single-ended signal for driving the spectrum analyzer.

The measurement principle is as follows. The noise source toggles between a hot and cold state. In the cold state, it produces a noise power of kTB and in the hot state, it produces a noise power of ENRkTB, where ENR is termed the *excess noise power ratio*. At the system output, the noise figure can be determined by comparing the output noise for the two source temperatures. The *output noise power ratio* is related to the ENR by

$$\mathrm{OPR} = \frac{\mathrm{ENR}kTBG_a + (F-1)kTBG_a}{FkTBG_a} = \frac{\mathrm{ENR} + (F-1)}{F}. \quad \text{(D.1)}$$

Solving for F yields

$$F = \frac{\mathrm{ENR} - 1}{\mathrm{OPR} - 1}. \quad \text{(D.2)}$$

The OPR can easily be measured using the spectrum analyzer.

In addition to noise figure measurements, an IP3 measurement can be performed with a very similar setup, shown in Figure D.3. In this case, the receiver input is driven by two signal sources at slightly different frequencies. The frequencies are selected so that the two input tones and their IM3 products fall within the receiver passband. A simple power splitter combines the two signals and 10dB attenuators are used to prevent direct cross-modulation of the output stages of the frequency synthesizers. Thus, each signal source sees a 26dB attenuation from the output of the other signal source. The third order intermodulation products produced by the receiver can be observed directly with the spectrum analyzer. It is important to note that the specified linearity of the AD9630 buffers is such that they do not contribute significantly to the observed distortion.

Finally, the same board setup can be used for measurement of the receiver frequency response using the arrangement depicted in Figure D.4. In this test, the tracking generator of the HP8590B spectrum analyzer stimulates the receiver input. Because the DUT has frequency conversion built into it, the tracking generator output must first be upconverted to the desired input frequency band, centered at 1.57542GHz. This is accomplished using a frequency synthesizer and a Minicircuits ZFM-2 diode ring modulator. A variable attenuator allows for rapid adjustment of the receiver input power without having to re-align the tracking generator. The attenuator also attenuates any spurious outputs of the ZFM-2 mixer.

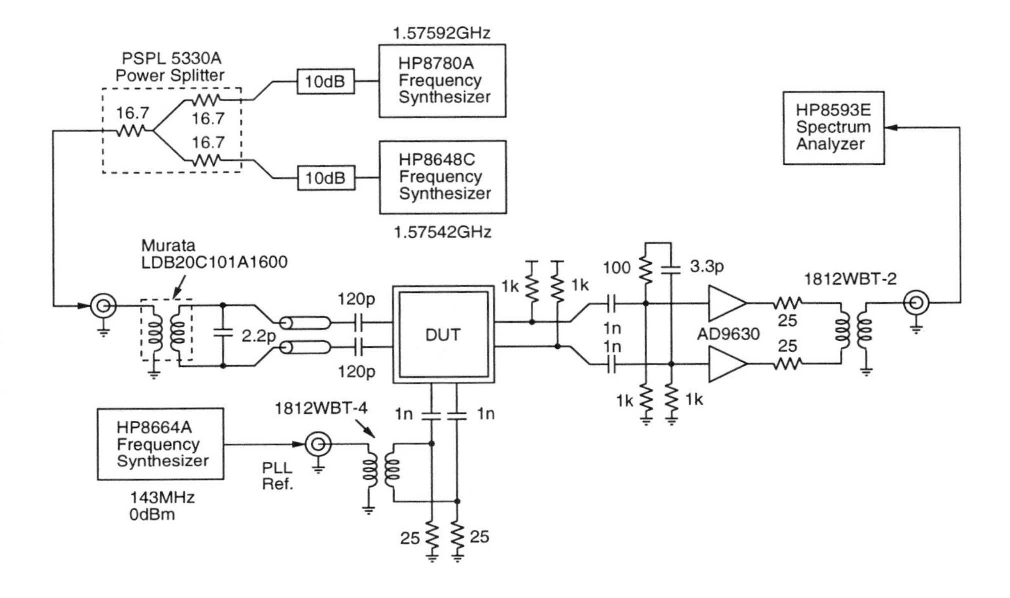

Figure D.3. Experimental setup for receiver IP3 measurements.

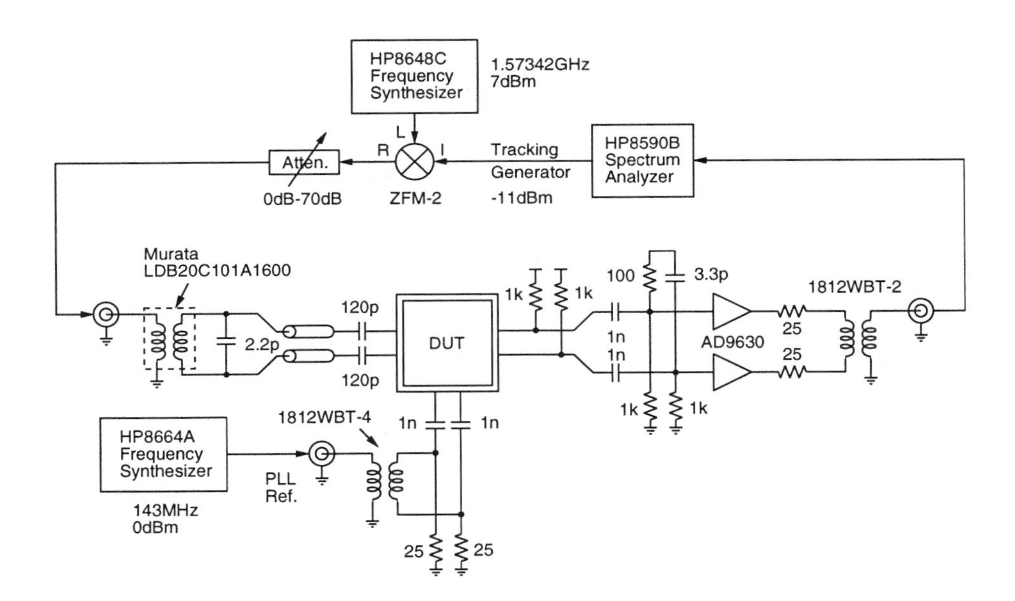

Figure D.4. Experimental setup for receiver frequency response measurements.

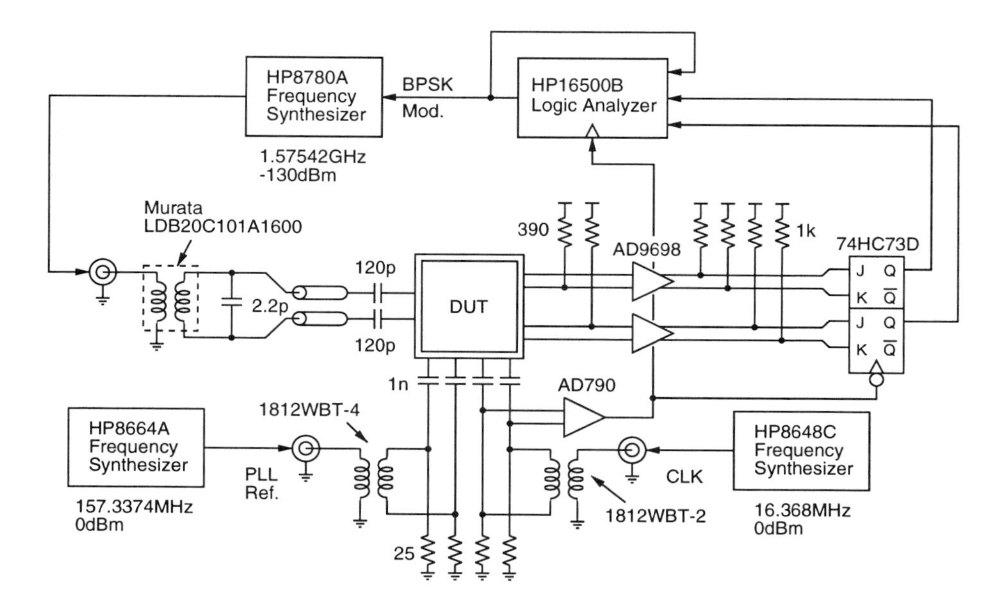

Figure D.5. Experimental setup for complete receiver measurements.

Once the tracking generator is properly aligned to the spectrum analyzer sweep oscillator, the frequency response of the DUT will be swept out and displayed on the screen. The dynamic range of frequency response measurements using this technique is limited to about 80dB by the spectrum analyzer.

3. WHOLE RECEIVER VERIFICATION

Finally, the entire receiver can be tested by applying a pseudo-noise modulated RF carrier to the input and acquiring the digital output bit streams from the I and Q channels. These bit streams can then be correlated by a computer to determine the SNR of the whole receiver. An experimental setup for doing such a measurement is shown in Figure D.5.

In this setup, the I and Q output channels are brought off-chip as low-swing differential signals to reduce coupling to the substrate. The signals are sensed with an AD9698 dual comparator and then latched into two J-K flip-flops. The comparators and flip-flops are clocked with the HP8648C frequency synthesizer, which also supplies the sampling clock to the DUT. This sampling clock is also a low-swing differential signal. An AD790 comparator amplifies this signal up to a full-swing logic level for operation of the comparators and flip-flops. The same clock signal also drives the logic analyzer, thereby ensuring that all parts of the system are on the same timebase. Note that all of the frequency synthesizers are slaved to the 10MHz crystal reference of the

HP8664A so that they also share the same timebase. This connection is omitted from the figure for clarity.

To operate this experiment, a pseudo-noise sequence is loaded into the pattern generator of the HP16500B logic analyzer. The pseudo-noise pattern modulates a 1.57542GHz carrier produced by the HP8780A frequency synthesizer. This modulated carrier stimulates the receiver input, and the resulting output data streams are acquired by the logic analyzer. The pseudo-noise code is also acquired by the analyzer to provide a convenient phase-reference for subsequent code correlations performed by computer.

The PLL reference frequency and the sampling CLK frequency are chosen to provide a sampling rate that is exactly sixteen times the pseudo-noise code rate of 1.023MHz and eight times the intermediate frequency of 2.046MHz. Note that a PLL reference frequency of 157.3374MHz produces an LO frequency of 1.573374GHz so that the IF frequency is 2.046MHz. This frequency plan causes pseudo-noise chips to occur at regular sample intervals, thereby simplifying the computational complexity in correlating the received IF signal with a reference pseudo-noise code. Even with such simplifications, the computation takes about 1 hour on a Sun Ultra 1 workstation.

References

[1] Aldert van der Ziel, *Noise in Solid State Devices and Circuits*, John Wiley & Sons, New York, 1986.

[2] A. A. Abidi, "High-frequency noise measurements on FET's with small dimensions," *IEEE Transactions on Electron Devices*, vol. ED–33, no. 11, pp. 1801–1805, Nov. 1986.

[3] Norman G. Einspruch, Ed., *VLSI Electronics: Microstructure Science*, vol. 18, chapter 1, pp. 1–37, Academic Press, New York, 1989.

[4] Bradford W. Parkinson, "Introduction and heritage of NAVSTAR, the global positioning system," In Parkinson and Spilker [104], pp. 3–28.

[5] Guglielmo Marconi, "Radio telegraphy," *Proceedings of the IRE*, vol. 10, pp. 215–238, Aug. 1922.

[6] Thomas H. Lee, *The Design of CMOS Radio Frequency Integrated Circuits*, Cambridge University Press, 1998.

[7] R. H. Mariott, "United States radio development," *Proceedings of the IRE*, vol. 5, pp. 179–198, June 1917.

[8] Guglielmo Marconi, "Radio communication," *Proceedings of the IRE*, vol. 16, pp. 40–69, Jan. 1928.

[9] Michael Riordan and Lillian Hoddeson, *Crystal Fire: The Birth of the Information Age*, W. W. Norton & Company, 1997.

[10] Lee de Forest, "The audion – detector and amplifier," *Proceedings of the IRE*, vol. 2, pp. 15–36, Mar. 1914.

[11] Haraden Pratt, "Long range reception with combined crystal detector and audion amplifier," *Proceedings of the IRE*, vol. 3, pp. 173–183, June 1915.

[12] Edwin H. Armstrong, "Operating features of the audion," *Electrical World*, , no. 24, pp. 1149–1152, Dec. 1914.

[13] John L. Hogan, Jr., "The heterodyne receiving system and notes on the recent Arlington-Salem tests," *Proceedings of the IRE*, vol. 1, pp. 75–97, July 1913.

[14] Benjamin Liebowitz, "The theory of heterodyne receivers," *Proceedings of the IRE*, vol. 3, pp. 185–204, Sept. 1915.

[15] Edwin H. Armstrong, "Some recent developments in the audion receiver," *Proceedings of the IRE*, vol. 3, pp. 215–247, Sept. 1915.

[16] Edwin H. Armstrong, "The super-heterodyne – its origin, development, and some recent improvements," *Proceedings of the IRE*, vol. 12, pp. 539–552, Oct. 1924.

[17] Edwin H. Armstrong, "A new system of short wave amplification," *Proceedings of the IRE*, vol. 9, pp. 3–27, Feb. 1921.

[18] W. G. Cady, "The piezo-electric resonator," *Proceedings of the IRE*, vol. 10, pp. 83–114, Jan. 1922.

[19] Stephen A. Maas, *Microwave Mixers*, Artech House, 1993.

[20] Edwin H. Armstrong, "Some recent developments of regenerative circuits," *Proceedings of the IRE*, vol. 10, pp. 244–260, Aug. 1922.

[21] D. G. Tucker, "The history of the homodyne and synchrodyne," *Journal of the British Institution of Radio Engineers*, vol. 14, no. 4, pp. 143–154, Apr. 1954.

[22] F. M. Colebrook, "Homodyne," *Wireless World and Radio Review*, vol. 13, pp. 645–648, 1924.

[23] H. de Bellescize, "La réception synchrone," *Onde Electronique*, vol. 11, pp. 209–272, 1932.

[24] Asad A. Abidi, "Direct-conversion radio transceivers for digital communications," *IEEE Journal of Solid-State Circuits*, vol. 30, no. 12, pp. 1399–1410, Dec. 1995.

[25] E. H. Colpitts and O. B. Blackwell, "Carrier current telephony and telegraphy," *Proceedings of the IRE*, vol. 40, pp. 205–300, 1921.

[26] R. A. Heising, "Modulation in radio telephony," *Proceedings of the IRE*, vol. 9, pp. 305–352, Aug. 1921.

[27] J. R. Carson, U.S. Patent #1343306, June 1920.

[28] J. R. Carson, U.S. Patent #1343307, June 1920.

[29] J. R. Carson, U.S. Patent #1449382, Mar. 1923.

[30] R. V. L. Hartley, "Relations of carrier and side-bands in radio transmission," *Proceedings of the IRE*, vol. 11, pp. 34–56, Feb. 1923.

[31] R. V. L. Hartley, U.S. Patent #1666206, Apr. 1928.

[32] Karl Willy Wagner, "Spulen- und kondensatorleitungen," *Archiv für Elektrotechnik*, vol. 8, July 1919.

[33] George A. Campbell, "Physical theory of the electric wave-filter," *Bell System Technical Journal*, vol. 1, no. 2, pp. 1–32, Nov. 1922.

[34] Otto J. Zobel, "Theory and design of uniform and composite electric wave-filters," *Bell System Technical Journal*, vol. 2, no. 1, pp. 1–46, Jan. 1923.

[35] D. K. Weaver, Jr., "Design of RC wide-band 90-degree phase-difference network," *Proceedings of the IRE*, vol. 42, no. 4, pp. 671–676, Apr. 1954.

[36] D.K. Weaver, Jr., "A third method of generation and detection of single-sideband signals," *Proceedings of the IRE*, pp. 1703–1705, June 1956.

[37] Leonard R. Kahn, "Single-sideband transmission by envelope elimination and restoration," *Proceedings of the IRE*, vol. 40, no. 7, pp. 803–806, July 1952.

[38] J. B. Johnson, "Thermal agitation of electricity in conductors," *Physical Review*, vol. 32, no. 7, pp. 97–109, July 1928.

[39] H. Nyquist, "Thermal agitation of electric charge in conductors," *Physical Review*, vol. 32, no. 7, pp. 110–113, July 1928.

[40] D. O. North, "The absolute sensitivity of radio receivers," *RCA Review*, vol. 6, pp. 332–343, Jan. 1942.

[41] H. T. Friis, "Noise figures of radio receivers," *Proceedings of the IRE*, vol. 32, no. 7, pp. 419–422, July 1944.

[42] H. Rothe and W. Dahlke, "Theory of noisy fourpoles," *Proceedings of the IRE*, vol. 44, no. 6, pp. 811–818, June 1956.

[43] H.A. Haus et al., "Representation of noise in linear twoports," *Proceedings of the IRE*, vol. 48, pp. 69–74, Jan. 1960.

[44] Keneth A. Simons, "The decibel relationships between amplifier distortion products," *Proceedings of the IEEE*, , no. 7, pp. 1071–1086, July 1970.

[45] A. Rofougaran et al., "A single-chip 900-MHz spread-spectrum wireless transceiver in 1-μm CMOS part II: Receiver design," *IEEE Journal of Solid-State Circuits*, vol. 33, no. 4, pp. 535–547, Apr. 1998.

[46] Barrie Gilbert, "A new wide-band amplifier technique," *IEEE Journal of Solid-State Circuits*, , no. 4, pp. 353–365, Dec. 1968.

[47] Barrie Gilbert, "A precise four-quadrant multiplier with subnanosecond response," *IEEE Journal of Solid-State Circuits*, , no. 4, pp. 365–373, Dec. 1968.

[48] Guillermo Gonzales, *Microwave Transistor Amplifiers: Analysis and Design*, Prentice Hall, 1984.

[49] Behzad Razavi, *RF Microelectronics*, Prentice Hall, 1998.

[50] Jan Craninckx and Michel S. J. Steyaert, "A 1.8-GHz CMOS low-phase-noise voltage-controlled oscillator with prescaler," *IEEE Journal of Solid-State Circuits*, vol. 30, no. 12, pp. 1474–1482, Dec. 1995.

[51] K. B. Ashby et al., "High Q inductors for wireless applications in a complementary silicon bipolar process," *IEEE Journal of Solid-State Circuits*, vol. 31, no. 1, pp. 4–9, Jan. 1996.

[52] C. Patrick Yue and S. Simon Wong, "On-chip spiral inductors with patterned ground shields for Si-based RF IC's," *IEEE Journal of Solid-State Circuits*, vol. 33, no. 5, pp. 743–752, May 1998.

[53] Yun-Ti Wang and A. A. Abidi, "CMOS active filter design at very high frequencies," In Tsividis and Voorman [97], pp. 258–269.

[54] David K. Su et al., "Experimental results and modeling techniques for substrate noise in mixed-signal integrated circuits," *IEEE Journal of Solid-State Circuits*, vol. 28, no. 4, pp. 420–430, Apr. 1993.

[55] Jan Crols and Michel S. J. Steyaert, "A single-chip 900 MHz CMOS receiver front end with a high performance low-IF topology," *IEEE Journal of Solid-State Circuits*, vol. 30, no. 12, pp. 1483–1492, Dec. 1995.

[56] J.C. Rudell et al., "A 1.9-GHz wide-band IF double conversion CMOS receiver for cordless telephone applications," *IEEE Journal of Solid-State Circuits*, vol. 32, no. 12, pp. 2071–2088, Dec. 1997.

[57] David H. Shen et al., "A 900-MHz RF front-end with integrated discrete-time filtering," *IEEE Journal of Solid-State Circuits*, vol. 31, no. 12, pp. 1945–1954, Dec. 1996.

[58] A.J. Van Dierendonck, "GPS receivers," In Parkinson and Spilker [104], pp. 329–407.

[59] Arvin R. Shahani et al., "Dividerless frequency synthesis using aperture phase detection," *IEEE Journal of Solid-State Circuits*, vol. 33, no. 12, Dec. 1998.

[60] R.P. Jindal, "Hot-electron effects on channel thermal noise in fine-line NMOS field-effect transistors," *IEEE Transactions on Electron Devices*, vol. ED–33, no. 9, pp. 1395–1397, Sept. 1986.

[61] Suharli Tedja, Jan Van der Spiegel, and Hugh H. Williams, "Analytical and experimental studies of thermal noise in MOSFET's," *IEEE Transactions on Electron Devices*, vol. 41, no. 11, pp. 2069–2075, Nov. 1994.

[62] Bing Wang, James R. Hellums, and Charles G. Sodini, "MOSFET thermal noise modeling for analog integrated circuits," *IEEE Journal of Solid-State Circuits*, vol. 29, no. 7, pp. 833–835, July 1994.

[63] Andrew N. Karanicolas, "A 2.7V 900MHz CMOS LNA and mixer," in *ISSCC Digest of Technical Papers*, 1996, vol. 39, pp. 50–51.

[64] A. Rofougaran et al., "A 1 GHz CMOS RF front-end IC for a direct-conversion wireless receiver," *IEEE Journal of Solid-State Circuits*, vol. 31, no. 7, pp. 880–889, July 1996.

[65] Samuel Sheng et al., "A low-power CMOS chipset for spread-spectrum communications," in *ISSCC Digest of Technical Papers*, 1996, vol. 39, pp. 346–347.

[66] J. Y.-C. Chang, A.A. Abidi, and M. Gaitan, "Large suspended inductors on silicon and their use in a 2-μm CMOS RF amplifier," *IEEE Electron Device Letters*, vol. 14, no. 5, pp. 246–248, May 1993.

[67] Young J. Shin and Klaas Bult, "An inductorless 900MHz RF low-noise amplifier in 0.9μm CMOS," in *Custom Integrated Circuits Conference Digest of Technical Papers*, 1997, pp. 513–516.

[68] R.R.J. Vanoppen et al., "RF noise modelling of 0.25μm CMOS and low power LNAs," in *International Electron Devices Meeting Digest of Technical Papers*, 1997, pp. 317–320.

[69] Qiuting Huang et al., "Broadband, 0.25μm CMOS LNAs with sub-2dB NF for GSM applications," in *Custom Integrated Circuits Conference Digest of Technical Papers*, 1998, pp. 67–70.

[70] Johan Janssens et al., "A 10mW inductorless, broadband CMOS low noise amplifier for 900 MHz wireless communications," in *Custom Integrated Circuits Conference Digest of Technical Papers*, 1998, pp. 75–78.

[71] R. Benton et al., "GaAs MMICs for an integrated GPS front-end," in *GaAs-IC Symposium Digest of Technical Papers*, 1992, pp. 123–126.

[72] Kenneth R. Cioffi, "Monolithic L-band amplifiers operating at milliwatt and sub-milliwatt DC power consumptions," in *IEEE Microwave and Millimeter-Wave Monolithic Circuits Symposium*, 1992, pp. 9–12.

[73] Masashi Nakatsugawa, Yo Yamaguchi, and Masahiro Muraguchi, "An L-band ultra low power consumption monolithic low noise amplifier," in *GaAs-IC Symposium Digest of Technical Papers*, 1993, pp. 45–48.

[74] E. Heaney et al., "Ultra low power low noise amplifiers for wireless communications," in *GaAs-IC Symposium Digest of Technical Papers*, 1993, pp. 49–51.

[75] Yuhki Imai, Masami Tokumitsu, and Akira Minakawa, "Design and performance of low-current GaAs MMIC's for L-band front-end applications," *IEEE Transactions on Microwave Theory and Techniques*, vol. 39, no. 2, pp. 209–215, Feb. 1991.

[76] N.H. Sheng et al., "A 30 GHz bandwidth AlGaAs-GaAs HBT direct-coupled feedback amplifier," *IEEE Microwave and Guided Wave Letters*, vol. 1, no. 8, pp. 208–210, Aug. 1991.

[77] Robert G. Meyer and William D. Mack, "A 1-GHz BiCMOS RF front-end IC," *IEEE Journal of Solid-State Circuits*, vol. 29, no. 3, pp. 350–355, Mar. 1994.

[78] Kevin W. Kobayashi and Aaron K. Oki, "A low-noise baseband 5-GHz direct-coupled HBT amplifier with common-base active input match," *IEEE Microwave and Guided Wave Letters*, vol. 4, no. 11, pp. 373–375, Nov. 1994.

[79] M. J. O. Strutt and A. van der Ziel, "The causes for the increase of the admittances of modern high-frequency amplifier tubes on short waves," *Proceedings of the IRE*, vol. 26, no. 8, pp. 1011–1032, Aug. 1938.

[80] Aldert van der Ziel, "Noise in solid-state devices and lasers," *Proceedings of the IEEE*, vol. 58, no. 8, pp. 1178–1206, Aug. 1970.

[81] R.P. Jindal, "Noise associated with distributed resistance of MOSFET gate structures in integrated circuits," *IEEE Transactions on Electron Devices*, vol. ED–31, no. 10, pp. 1505–1509, Oct. 1984.

[82] Behzad Razavi, Ran-Hong Yan, and Kwing F. Lee, "Impact of distributed gate resistance on the performance of MOS devices," *IEEE Transactions on Circuits and Systems – I: Fundamental Theory and Applications*, vol. 41, no. 11, pp. 750–754, Nov. 1994.

[83] R.P. Jindal, "Distributed substrate resistance noise in fine-line NMOS field-effect transistors," *IEEE Transactions on Electron Devices*, vol. ED–32, no. 11, pp. 2450–2453, Nov. 1985.

[84] Paul R. Gray and Robert G. Meyer, *Analysis and Design of Analog Integrated Circuits*, chapter 2, p. 116, John Wiley & Sons, Inc., New York, 1993.

[85] Aldert van der Ziel, "Gate noise in field effect transistors at moderately high frequencies," *Proceedings of the IEEE*, pp. 461–467, Mar. 1963.

[86] François M. Klaassen and Jan Prins, "Noise of field-effect transistors at very high frequencies," *IEEE Transactions on Electron Devices*, vol. ED–16, no. 11, pp. 952–957, Nov. 1969.

[87] Gleb V. Klimovitch, Thomas H. Lee, and Yoshihisa Yamamoto, "Physical modeling of enhanced high-frequency drain and gate current noise in short-chanel MOSFETs," in *First International Workshop on Design of Mixed-Mode Integrated Circuits and Applications*, 1997, pp. 53–56.

[88] Ping K. Ko, Chenming Hu, et al., *BSIM3v3 Manual*, Department of Electrical Engineering and Computer Science, University of California, Berkeley, 1995.

[89] Bing Wang, "Wide band noise in MOSFETs," M.S. thesis, Massachusetts Institute of Technology, Oct. 1992.

[90] Christopher D. Hull and Robert G. Meyer, "A systematic approach to the analysis of noise in mixers," *IEEE Transactions on Circuits and Systems – I: Fundamental Theory and Applications*, vol. 40, no. 12, pp. 909–919, Dec. 1993.

[91] H. P. Walker, "Sources of intermodulation in diode-ring mixers," *The Radio and Electronic Engineer*, vol. 46, no. 5, pp. 247–255, May 1976.

[92] J. G. Gardiner, "The relationship between cross-modulation and intermodulation distortions in the double-balanced modulator," *Proceedings of the IEEE*, vol. 56, no. 11, pp. 2069–2071, Nov. 1968.

[93] N. M. Nguyen and R. G. Meyer, "Si IC-compatible inductors and LC passive filters," *IEEE Journal of Solid-State Circuits*, vol. SC-25, no. 4, pp. 1028–1031, Aug. 1990.

[94] Clark T.-C. Nguyen, "Microelectromechanical devices for wireless communications," in *Proceedings of the 1998 IEEE International Micro Electro Mechanical Systems Workshop*, Jan. 1998, pp. 1–7.

[95] R. P. Sallen and E. L. Key, "A practical method of designing RC active filters," *IRE Transactions*, vol. CT–2, no. 3, pp. 74–85, Mar. 1955.

[96] J. O. Voorman, "Continuous-time analog integrated filters," In Tsividis and Voorman [97], pp. 15–46.

[97] Y. P. Tsividis and J. O. Voorman, Eds., *Integrated Continuous-Time Filters: Principles, Design and Applications*, IEEE Press, 1993.

[98] Gert Groenewold, "The design of high dynamic range continuous-time integratable bandpass filters," *IEEE Transactions on Circuits and Systems – I: Fundamental Theory and Applications*, vol. 38, no. 8, pp. 838–852, Aug. 1991.

[99] H. Khorramabadi and P.R. Gray, "High-frequency CMOS continuous-time filters," In Tsividis and Voorman [97], pp. 221–230.

[100] Rajesh H. Zele and David J. Allstot, "Low-power CMOS continuous-time filters," *IEEE Journal of Solid-State Circuits*, vol. 31, no. 2, pp. 157–168, Feb. 1996.

[101] François Krummenacher and Norbert Joehl, "A 4-MHz CMOS continuous-time filter with on-chip automatic tuning," *IEEE Journal of Solid-State Circuits*, vol. 23, no. 3, pp. 750–758, June 1988.

[102] Akira Yukawa, "A CMOS 8-bit high-speed A/D converter IC," *IEEE Journal of Solid-State Circuits*, vol. SC–20, no. 3, pp. 775–779, June 1985.

[103] C.P. Yue, C. Ryu, J. Lau, T.H. Lee, and S.S. Wong, "A physical model for planar spiral inductors on silicon," in *1996 International Electron Devices Meeting Digest of Technical Papers*, Dec. 1996, pp. 155–158.

[104] Bradford W. Parkinson and James J. Spilker, Jr., Eds., *Global Positioning System: Theory and Applications*, vol. I, American Institute of Aeronautics and Astronautics, Inc., 1996.

Index

About the Authors

Derek K. Shaeffer received his B.S. degree from the University of Southern California in 1993, and his M.S. and Ph.D. degrees from Stanford University in 1995 and 1998. From 1992-1997, he worked for Tektronix, Inc. in Beaverton, Oregon where he cut his teeth designing A/D converter and communications circuits in CMOS and bipolar technologies. His current research interests are in CMOS and bipolar implementations of low noise, high linearity wireless communications receivers.

In his spare time, he enjoys playing piano and writing music for the piano and guitar.

Thomas H. Lee received the S.B., S.M. and Sc.D. degrees in electrical engineering, all from the Massachusetts Institute of Technology in 1983, 1985, and 1990, respectively.

He joined Analog Devices in 1990 where he was primarily engaged in the design of high-speed clock recovery devices. In 1992, he joined Rambus Inc. in Mountain View, CA where he developed high-speed analog circuitry for 500 megabyte/s CMOS DRAMs. He has also contributed to the development of PLLs in the StrongARM, Alpha and K6/K7 microprocessors. Since 1994, he has been an Assistant Professor of Electrical Engineering at Stanford University where his research interests are in low-power, high-speed analog circuits and systems, with a focus on gigahertz-speed wireless integrated circuits built in conventional silicon technologies, particularly CMOS.

He has twice received the "Best Paper" award at the International Solid-State Circuits Conference, was co-author of a "Best Student Paper" at ISSCC, recently won a Packard Foundation Fellowship, and is a Distinguished Lecturer of the IEEE Solid-State Circuits Society. He holds six U.S. patents and is the author of a textbook, *The Design of CMOS Radio-Frequency Integrated Circuits*, Cambridge Press, 1998, and a co-author of two additional books on RF circuit design.